T0292105

Cambridge Studies in Historical Geography 24

MARC BLOCH, SOCIOLOGY AND GEOGRAPHY

Cambridge Studies in Historical Geography 24

Series editors:
ALAN R. H. BAKER, RICHARD DENNIS, DERYCK HOLDSWORTH

Cambridge Studies in Historical Geography encourages exploration of the philosophies, methodologies and techniques of historical geography and publishes the results of new research within all branches of the subject. It endeavours to secure the marriage of traditional scholarship with innovative approaches to problems and to sources, aiming in this way to provide a focus for the discipline and to contribute towards its development. The series is an international forum for publication in historical geography which also promotes contact with workers in cognate disciplines.

For a full list of titles in the series, please see end of book.

MARC BLOCH, SOCIOLOGY AND GEOGRAPHY

Encountering changing disciplines

SUSAN W. FRIEDMAN

Pennsylvania State University

CAMBRIDGE
UNIVERSITY PRESS

PUBLISHED BY THE PRESS SYNDICATE OF THE UNIVERSITY OF CAMBRIDGE
The Pitt Building, Trumpington Street, Cambridge, United Kingdom

CAMBRIDGE UNIVERSITY PRESS
The Edinburgh Building, Cambridge CB2 2RU, UK
40 West 20th Street, New York NY 10011–4211, USA
477 Williamstown Road, Port Melbourne, VIC 3207, Australia
Ruiz de Alarcón 13, 28014 Madrid, Spain
Dock House, The Waterfront, Cape Town 8001, South Africa

http://www.cambridge.org

First published 1996
First paperback edition 2004

A catalogue record for this book is available from the British Library

Library of Congress cataloguing in publication data

Friedman, Susan W.
    Marc Bloch, sociology and geography : encountering changing
disciplines / Susan W. Friedman.
        p.   cm.
    ISBN 0 521 56157 4 (hardback)
    1. Bloch, Marc Léopold Benjamin, 1886–1944.   2. Social sciences
and history.   3. Historical geography.   I. Title.
    D15.B596F75   1996
    304.2′3–dc20 95-37598 CIP

ISBN 0 521 56157 4 hardback
ISBN 0 521 61215 2 paperback

Transferred to digital printing 2004

Contents

Tables

Acknowledgments

A number of individuals graciously granted me permission to consult and cite from various collections of private papers that were critical for this work. For access to and permission to cite from the correspondence between Marc Bloch and Lucien Febvre, I would like to thank both Etienne Bloch and the late Fernand Braudel. Etienne Bloch also gave me permission to use and cite from the Marc Bloch papers held at the Archives Nationales, as well as from various other sets of papers which he holds personally, and I am deeply grateful for all he has done to facilitate my research. Pierre Gasnault, the chief librarian of the Bibiliothèque Mazarine, kindly allowed me to use and cite from the Fonds Demangeon-Perpillou and also provided valuable assistance in the dating and identification of the correspondence held in that collection. I am also grateful to Monsieur le Comte Jacques-Henri Pirenne not only for his permission to use and cite from the archives of the Pirenne family, but also for his hospitality while I was working at the family château in Hierges. In Toronto, Father Laurence K. Shook provided assistance with the papers of Etienne Gilson held by the Pontifical Institute of Mediæval Studies.

This book is an outgrowth of a doctoral dissertation for the Department of Geography at the University of Toronto. The late Howard F. Andrews supervised that work and his knowledge and critical eye contributed enormously to my research. Without his enthusiastic support, I would have never embarked on this project, and our many long conversations helped to shape my work in innumerable ways. Among the many others at the University of Toronto whom I would like to thank are the late Joe May, Irving Zeitlin, Jock Galloway, Gunter Gad, Aidan McQuillan, Ted Relph, Jim Lemon, Cecil Houston, and Zehra Alpar. I am also very grateful for the subsequent encouragement and advice from Father Ambrose Raftis, Brian Stock, and Miriam Brokaw. The counsel and assistance of Jacques Revel added greatly, particularly in the early stages of my research. In addition I would like to thank Clémens Heller, Philippe Besnard, Marie-Claire Robic, Catherine Rhein, John E. Craig and Marleen Wessel for their help and assistance at certain critical stages in my research, Larry Kerslake, Harold Slamovitz, and Vera Mark for assistance with some problems of translation and expression, Tina Jacquette for divulging some of the mysteries of word processing, and Rodney Erickson for his patience on other mutual projects. I am very indebted to innumerable librarians and archivists and of these I would especially like to acknowledge the patience and assistance of those at the microfilm room of the Archives Nationales in Paris and the interlibrary loan rooms of the Robarts Library at the University of Toronto and the Pattee Library at the Pennsylvania State University. More generally, the extensive collections of the Robarts Library and the financial support provided by the University of Toronto, Ontario Graduate and Connaught Scholarships were invaluable in the formative stages of my research. Subsequently, able advice from Ruth Parr and Sheila Kane of Cambridge University Press and from the anonymous referees helped to substantially improve the manuscript.

On a more personal note, thanks must go to Pauline Chan, Adrian Shaw, and my office mates at Toronto (Mark Schroeder, Dick Morino, and Murdo McPherson) and at Penn State (Sam Lowe) whose friendship and collegial insights have helped me along the way, to my parents for instilling in me a love of books and languages, and to my children, Peter and Pauline, for helping me to put everything into perspective. Finally, I would particularly like to thank Deryck Holdsworth whose insights, criticism, and encouragement have seen me through this entire project. Without his unfailing support, this book would probably never have been completed.

A note on translation

Marc Bloch was keenly interested in the changing meaning of words and as a result his work can be difficult to translate. This work explores the meanings which he invested in certain terms and how those meanings differed at times from those which others might associate with the same terms. Existing translations of Bloch's work have attempted to put his prose, which can be rather dense at times, into flowing English, but in so doing have often sacrificed too much of his meaning for my purposes. As a result, the translations in this book are my own. I have tried to stay as close to the original meaning as possible both in my choice of words and in style. Accordingly, I have retained many of Bloch's qualifications and modifications, which can make reading a bit difficult but does give a more accurate portrayal of his thought. When faced with a choice between elegance of expression and retention of meaning, I have chosen the latter.

Abbreviations

Abbreviations used when citing journals are as follows:

AESC	*Annales: Economies, Sociétés, Civilisations*
AG	*Annales de Géographie*
AHES	*Annales d'Histoire Economique et Sociale*
AHS	*Annales d'Histoire Sociale*
AS	*Année Sociologique*
BFLS	*Bulletin de la Faculté des Lettres de Strasbourg*
BSHM	*Bulletin de la Société d'Histoire Moderne*
MHS	*Mélanges d'Histoire Sociale*
NCSS	*Notes Critiques: Sciences Sociales*
RFS	*Revue Française de Sociologie*
RH	*Revue Historique*
RIE	*Revue Internationale de l'Enseignement*
RS	*Revue de Synthèse*
RSH	*Revue de Synthèse Historique*
RU	*Revue Universitaire*

Abbreviations for archival sources are as follows:

AN   Papers held at the Archives Nationales in Paris, followed by their number. For the microfilms, the page numbers used on the film are also given.

EB   Papers held by Etienne Bloch, La Haye, France.

DP   Le Fonds Demangeon-Perpillou, held at the Bibliothèque Mazarine in Paris.

HP   The Pirenne Archives, formerly at Hierges, France, now at the Université Libre de Bruxelles.

EG   Etienne Gilson papers, Pontifical Institute of Medieval Studies, Toronto, Canada.

Introduction

When preparing Marc Bloch's unfinished *Apologie pour l'Histoire* for publication in 1949, Lucien Febvre found the following observation:

each science, taken separately, often finds its most successful artisans in renegades from neighboring sectors. Pasteur, who renewed biology, was not a biologist – and during his lifetime, one made him see that clearly; just as Durkheim and Vidal de la Blache, who left an incomparably deeper mark than any specialist on historical studies at the beginning of the twentieth century, did not rank among the certified historians, the first being a philosopher turned sociologist and the second a geographer.[1]

Bloch himself was one of those historians who began their training at the beginning of the twentieth century when Emile Durkheim and Paul Vidal de la Blache had such an impact. In 1931, when his generation was well established in the historical profession, he still felt that history had much to learn from both geography and sociology. These fields were "the two great disciplines of which one can say that they renewed or ought to have renewed history."[2]

Despite the claims of many scholars, Bloch did not adopt either the Durkheimian or Vidalian approaches, which so intrigued him, and would never refer to himself in those terms. When he was in his mid-twenties and working toward his doctorate, he had already questioned the usefulness to history of the regional method so championed at that time by the Vidalians as the touchstone of their approach. Similarly, he dismissed one of the central concepts of contemporary geographical writings when he wrote, "In sum the '*pays*' rarely offers a useful framework for historical research."[3] In early reviews without referring directly to sociology, Bloch also warned historians against overly theoretical and doctrinal approaches which could obscure "the reality" of history.[4]

Bloch's relationships to the fields of geography and sociology, in which he held such hopes for the revitalization of history, were fluctuating and at times uneasy. He adapted and in the process modified what he found most intriguing

for his own work on the social history of France; at the same time he continued to criticize the work produced within geography and sociology. His perspectives on these fields are examined together here not only because both are
central to his approach but also because what he took from one was colored
by what he took from the other.

Although many scholars have been attracted to Bloch's work, the message
that they convey is a confusing and even contradictory one. Some have simply
praised its interdisciplinary character without examining how Bloch dealt
with the sometimes conflicting aims and methods of the disciplines on which
he drew. Others, including in recent years a number of both historical sociologists and historical geographers, have cited Bloch as a precedent for their own
particular approach, be it analytic or interpretive, model-based or concerned
with an understanding of the particular. Only by studying his changing relationships to both geography and sociology can one begin to understand the
tension in his work revealed indirectly in the differing interpretations that it
has received. Informed by social science and yet still very much interested in
differences rather than laws, Bloch's work does not fit neatly into the categor-
izations which some have implied.

Another source of confusion over the character of Bloch's approach has
been the tendency of some to equate it with a perceived approach identified
with the "Annales school of history." This practice has helped to obscure the
significant criticisms which Bloch made of geography. In order to understand
Bloch's relationships to the field of geography, one must look at that work
directly, not through the lenses of Lucien Febvre and Fernand Braudel.
Although Bloch was a co-founder of the *Annales*, his approach differed significantly from that of Lucien Febvre, who was much closer in spirit to the
Vidalians and who was also Braudel's mentor.[5]

It will not be enough, however, just to examine Bloch's writings on geography and sociology. The French historians, geographers, and sociologists of this
period did not write in a vacuum. In the centralized world of French educational institutions, many references were implied with the assumption that all
interested would understand. To bridge the cultural and historical gap presented to present-day readers outside France, one must try to reconstruct some
of that world in order to begin to understand what was said and what was not.
The aim here is not to provide a reductionist explanation of Bloch's views but
instead to place them in a setting in which they can be understood, to clarify
the terms of the debates, and to suggest not only what was common in the
respective writings but also what was different. Only in this way can Bloch's
particular contribution really be understood. This process of reconstruction
cannot be limited to an examination of the academic writing done by Bloch
and other contemporaries but must also include the institutional setting and
constraints within which they worked, for it is at such a level that the possibilities and difficulties of interdisciplinary work become most apparent.[6]

Here the context for examining Bloch's work will be predominantly French and predominantly academic. To understand Bloch, who, for his times, had an exceptionally broad international orientation, a closer examination of the international context would also be useful. However, given the complexities involved in examining three separate disciplines in France and the transferences between them, it was felt best to leave such discussions for now.[7] Similarly, though some allusions are made to the ideologies and political underpinnings informing Bloch's work and that of his contemporaries, I have left a more detailed scrutiny of such issues to others.

Bloch entered the University of Paris when the intellectual changes associated with what was known as the Nouvelle Sorbonne were at a high point. Representatives of established disciplines, such as Bloch's field of history, struggled to make their fields more scientific and those representing the relatively new social sciences, such as human geography and sociology, attempted to establish separate identities for theirs. As new fields and subfields proliferated, a number of academics and educational administrators pushed for interdisciplinary communication, and various spokesmen suggested that their particular discipline was best suited to bring the many fields together. Despite this dynamic intellectual setting, institutional inertia meant that students such as Bloch were subjected to academic programs which had changed little.

As a student in Paris at this time, Bloch was well placed both to understand the academic world of his father's generation and to witness the unresolved debates over new disciplinary definitions and approaches. Then, as a professor at the post-war University of Strasbourg, he was once again at the center of attempts to restructure the university and to explore interdisciplinary links. From his perspective one can view the tensions resulting from attempts to professionalize academic disciplines and the questions posed by the emergence of social science. More specifically, for one interested in the possible relationships between the fields of history, geography, and sociology, he is ideally situated.

Bloch's own position on these questions unfolded slowly. As a budding French historian at the turn of the century, he was not predisposed to methodological argument and took some time to articulate his views. They are not to be found in a treatise on the topic, but instead scattered throughout his published and unpublished writings, correspondence, and personal files. Exposed to the debates over the status of history, sociology, and geography as a student, he returned to them again and again – sometimes by substantive example and sometimes addressing the questions more directly. Although adopting some of the key concepts and catch phrases of his teachers, such as "collective representations" and *genre de vie*, he eventually invested them with new meanings. This translation made Bloch's approach somewhat more palatable to the historical profession, even if still controversial at times.[8] To understand this process, one must first understand what the terms meant for those who popularized them. Similarly, only with some understanding of the questions which troubled his teachers can one understand why he chose to address certain methodological issues and not others and how his answers differed from theirs.

The issues that preoccupied Bloch and his teachers included those of disciplinary boundaries and definition, the meaning of science, and the appropriate form of explanation. Sociologists argued that their field was *the* science of social facts, and geographers suggested a number of objects of study for theirs, including those of region, landscape, places, and the relationships between man and his environment. Historians, on the other hand, continued to depict their field as simply the study of the past, although they would argue amongst themselves whether it was past events or social phenomena that were most important. Arguments over the meaning of science touched on such issues as whether a science must seek to establish laws, whether explanation by final causes was admissible, and whether one could explain one order of phenomena by another – as in the efforts of some to explain social phenomena by either individual or physical ones. Other arguments over the appropriate form of explanation included those of the virtues and dangers of the comparative, chronological, and regional approaches.

Bloch's academic career can be seen as falling into two periods. The first, which he himself termed his "years of apprenticeship,"[9] included his years of training in Paris and teaching in the lycées of Montpellier and Amiens, followed by the disruption of the First World War. This period extends from 1904 when he entered the Ecole Normale Supérieure to his thesis defense in 1920 (the subject of Part I). During these years, Bloch was rather restricted in the kinds of writing that he could do. Following the war, when he received a teaching post at the University of Strasbourg, Bloch had more opportunity to address the questions that he found most interesting and was able to become a much more active practitioner and critic (the subject of Part II). Before turning to an examination of Bloch's apprenticeship years, I will give a brief overview of his education and career, situating him within the structures of the French Université – structures which not only helped to shape the paths he took but also established the limits against which he struggled.

Intended as a contribution to the history of thought in the social sciences, the goal of this book is to deepen our understanding of how Bloch met the challenges of the fields of geography and sociology, which he viewed as so important to the revitalization of history, or even more broadly, "human studies." As somewhat of a "renegade from a neighboring sector" himself, he can offer some interesting insights into the strengths and weaknesses of these two fields during the critical phase of their emergence in France – fields which continue to question both their own identities and their relationships with other disiplines and whose status in the university remains unclear.

1

Marc Bloch and the "Université"

Marc Bloch was a child, member, and critic of the French "Université." This unique and highly centralized state corporation, dating from Napoleon, included teaching posts from the primary to post-graduate levels, state-run inspections, teaching evaluations, examinations, and degrees. Born into an intensely academic family at the summit of the Université, Bloch attended its most prestigious institutions. Within its bounds, he grappled with questions of disciplinary development, methodology, and limits, including, for him, the central questions of the possible relationships between history, geography, and sociology. Increasingly, as his career advanced, he openly criticized its institutional structures which he found so confining and counterproductive. In brief, the Université provided the framework for both his life and his work, setting limits, charting obligatory paths, and providing the structures which he and others sought to change.

The seats of power in the Université were in Paris, and it was there that Marc Bloch spent his formative years. His father, Gustave Bloch, was appointed to the post of *maître de conférences* at the Ecole Normale Supérieure in 1887, the year after Marc's birth in Lyon. With the exception of two years away for military service and a year of study in Germany, he stayed in Paris until the age of 27 when he began his first position teaching in a provincial lycée.

Turn-of-the-century Paris was marked by the excitement, confusion, and ambivalence associated with rapid modernization and nationalization. In the eyes of some, electricity, improved sanitation, the Métro, and the spectacular shows at the Théatre de Châtelet and the Paris Opéra all served to make Paris a city of lights in contrast to the dreary, backward, and parochial provinces. And yet Paris could also be seen as a place of depravity where no values were sacred; where crime, narcotics, the occult, decadent sexual practices, and suspect foreign influences all reigned unchecked. To many, despite what were sometimes seen as dangerous urban influences, Paris was the place of intellect and achievement – a notion reinforced by the structure of the civil service, including the Université. Junior members were relegated to the small provincial towns and only gradually promoted to larger ones. If very successful, they might eventually reach Paris.

Many provincials both envied and resented the power and prestige of the capital, and a number of regionalist movements sought to address the imbalance. Some, in places such as Brittany, were separatist movements, but others, such as the Fédération régionaliste française founded in 1900, aimed more at a conservation of local life and initiative within a unified France – a goal to be achieved by such means as restructuring of the national educational system. Suggested changes included greater attention to local and regional concerns at all levels of the Université, a less centralized administrative system, and incentives for local instructors at all levels to remain in their region of origin.[1]

For France as a whole, it was a rather unsettling period. Following decades of political instability, the Third Republic proved remarkably resilient and yet was constantly challenged by crises including the Boulangist movement, the Panama scandal, and the Dreyfus Affair. Many felt that it could fall at any time. The Universal Exposition of 1900, intended as a celebration of the progress of civilisation – particularly French – during the nineteenth century, demonstrated instead the industrial and commercial superiority of Germany, England, and the United States.[2] To make matters worse, at the 1904 Congress of Arts and Sciences held in conjunction with the Louisiana Purchase Exhibition, German scholars dominated many fields.

Of all the crises, it was the Dreyfus affair which affected the young Marc Bloch most directly. Marc was only eight when Alfred Dreyfus was convicted of treason and only eleven when Emile Zola published his famous indictment of the army for making a scapegoat out of Dreyfus. Nevertheless, he later identified himself with the generation of the Dreyfus Affair.[3] The "Affair" had a profound impact on turn-of-the-century France and more specifically on the tone of the Ecole Normale Supérieure, where Bloch was a student. In the divisive debates, Dreyfusard intellectuals were pitted against anti-Semitic, monarchist, and right-wing Catholic forces, and existing schisms in French society were reinforced. As the arguments of the Dreyfusards gained force, the ties between church and state were severed and there were attempts to republicanize the army. The secular world of the Université became even more distinct and prestigious as a result, including changes not just at a national level but also in small towns where state teachers increasingly took on local leadership roles.[4]

Like Dreyfus, Bloch came from a family of highly assimilated French Jews with roots in Alsace and a tradition of French patriotism. In 1783, a great-grandfather had fought for France against Prussia and in 1870, Gustave Bloch had fought in defence of Strasbourg. Gustave Bloch was also among the professors at the Ecole Normale Supérieure who became active in the Dreyfusard movement, an involvement of which his son was keenly aware in light of discussions which took place in his home.[5] The young Marc clearly supported the Dreyfusard cause and the values which it represented.

In addition to being highly assimilated, patriotic and republican, Bloch's family was very academic. According to the standards set by the Université, his father, Gustave Bloch, cut a rather imposing figure. Teaching evaluations in the years before he was called to Paris describe him as a distinguished scholar and excellent professor but also as rigid and hard to please. An official report on his thesis defence in 1884 describes him as exceptionally mature and strong, sure of himself, vigorous, and well spoken.[6] To his students at the Ecole Normale Supérieure he was known as the "Méga," a name also associated with a skeleton of a megatherium housed at the school. Although he proceeded slowly in lectures, students found his analyses to be extremely perceptive, clear, and valuable. His correction of student lessons could be cutting, severe, and to the point. In the words of Raoul Blanchard (*promotion* [entering class of] 1897), "it was splendid and terrifying."[7]

Marc Bloch had only one sibling, Louis, who was seven years older. On Louis's premature death in 1922, Marc wrote of "the close intellectual community" he had had with his brother.[8] However, not unlike his father, Louis was a rather intimidating role model for the young Marc. According to Jérome Carcopino, Gustave adored his oldest son who was "endowed with all the aptitudes, curious about all the novelties, sensitive to all forms of intellectual activity, scholarly and artistic, enthusiastic and level-headed, thoughtful and cheerful."[9] Describing his first encounter with Marc Bloch at the Blochs' home in 1902, Lucien Febvre later wrote, "From this fleeting meeting, I have kept the memory of a slender adolescent with eyes brilliant with intelligence and timid cheeks – a little lost then in the radiance of his older brother, future doctor of great prestige."[10] Before his premature death, Louis became the laboratory chief of the diphtheria section of the Hôpital des Enfants Malades.[11]

The family home for about twenty years was at 72, Rue d'Alésia in the 14th arrondissement. It was not a particularly religious household since both Gustave and Sarah Bloch were highly assimilated Jews, but was undoubtedly an intensely academic one. Sarah dedicated herself to her husband's career and her sons' education, and Gustave began Marc's historical training at a very young age.[12] When Marc entered the Ecole Normale in 1904, Gustave became one of his teachers there as well. Shortly after his father's death in 1923, Marc testified, "I owe the better part of my training as an historian to my father; his lessons, begun in childhood and which never ceased, have marked me with what I hope is an ineffaceable imprint."[13]

Marc Bloch's academic credentials were impeccable. He attended the prestigious Lycée Louis-le-Grand for three years, ranking at the top and winning first prizes in history, French, English, Latin, and natural history. The lycée, a secondary school owned, run, and fully funded by the state, formed the core and pride of the Université in late nineteenth-century France and remained central to it in the twentieth century. It contrasted with both the municipal "collèges" which relied on local funding and the various church supported secondary schools. Of all the lycées, those in Paris were the most prestigious and as such viewed as stepping stones to the better institutions of higher education. In particular, Lycée Louis-le-Grand and Lycée Henri IV were favored to prepare for the Ecole Normale Supérieure.[14] In 1903, Bloch passed the *baccalauréat* in Letters and Philosophy (a school-leaving certificate) with the distinction "très bien," and in 1904, he was awarded a scholarship to the Ecole Normale Supérieure after passing the entrance examination with high marks in history.[15]

A unique institution, the Ecole Normale Supérieure (ENS) had been intially designed as a very select teachers' college, but had come to serve as a miniature elite faculty of letters training not only lycée and faculty professors but also politicians and journalists. Once at the ENS, where he specialized in medieval history, Bloch continued to impress even though his teachers noted that some of his lessons were a little bit dense and complicated and that at times he chose overly ambitious subjects.[16]

Bloch's next major hurdle was the *agrégation*. Held annually in Paris, this lengthy national examination served to select teachers for the better lycée posts and some faculty positions. The demanding examination was given over a period of three weeks including both written and oral components, and had a failure rate as high as 70 to 80 percent. ENS students (known as *normaliens*) had a considerable advantage, in part due to the compositions of the juries, which often included their own professors. In addition, provincial students lacked the library resources of their *normalien* counterparts, and many were already burdened with teaching responsibilities. As a result, some would take and retake the examination over a period of years before either receiving the title of *agrégé* and access to better teaching positions or resigning themselves to lower posts within the Université.[17] Not only did a much higher percentage of *normaliens* acquire the title of *agrégé*, they also took a much shorter time to do so.

History and geography shared an *agrégation*, which Bloch passed, gaining second place in 1908.[18] In place since 1830,[19] this joint *agrégation* both helped and limited geography's status. Tied to one of the established subjects, it was guaranteed representation, but also given an auxiliary role. As geography became more established as a professional discipline, some geographers, such as Ludovic Drapeyron and later Emmanuel de Martonne, pushed for an independent *agrégation* for geography, but this was not achieved until 1944.[20] In any case, this *agrégation* insured that Bloch, like his historical colleagues, would be very familiar with the changing discipline of geography.

After becoming an *agrégé*, Bloch studied for a year at Berlin and Leipzig. Academic pilgrimages of promising *agrégés* to Germany were an integral part of the professionalization of the French historical profession during the late nineteenth and early twentieth centuries. Begun under Victor Duruy (then

Minister of Public Instruction) in the 1860s, these excursions exposed the young historians to German scholarship and their seminar system which contrasted greatly with the rhetoric approach common in nineteenth-century France, and so were viewed as models. Following the Franco-Prussian War, the young scholars who studied in Germany began to adopt a somewhat more critical attitude toward the German university system, but still found much to admire. The historians who went in these early years would later be key actors in the renewal of French history, including Gaston Paris, Gabriel Monod, and Charles Seignobos.[21]

For Bloch, his year in Germany in 1908 provided an opportunity to study with Max Sering and Rudolph Eberstadt in Berlin and with Karl Bücher in Leipzig. The University of Leipzig was also well known for its interdisciplinary exchanges. Until shortly before Bloch's arrival, the historian Karl Lamprecht had worked closely with the geographer Friedrich Ratzel and the psychologist Wilhelm Wundt. This tradition was carried on in 1906 in the Seminar für Landesgeschichte and Siedlungskunde run by one of Lamprecht's students, Rudolf Kötzschke – a seminar which was focussed on rural history and as such closely tied to Bloch's developing interests.[22]

In 1909, Bloch entered the highly selective Fondation Thiers as one of a very small group of doctoral candidates who shared a residence and were given financial support to allow them to devote their time to their dissertations. Just as *normaliens* had a distinct advantage in taking the *agrégation*, the *pensionnaires* at the Fondation Thiers (as well as of a few other select institutions such as the Ecole Française d'Athènes) were given the time and support to get their doctorates well under way, a task others would have to fit in along with highly prescribed teaching duties. This difference translated into years of waiting and a much smaller chance at the more prestigious posts.[23] In 1912, Bloch began his stint of teaching in the provinces with a one-year appointment to the lycée in Montpellier, and the following year he taught at the lycée in Amiens.

Intensely patriotic, Bloch felt it was an honor and duty to fight for his country. Between 1905 and 1906 he had taken a year's leave from the ENS to fulfill the requirement for military service and following the outbreak of war in 1914, he entered the army as a sergeant. He experienced trench warfare with periods of idleness and battle, was wounded, had a serious bout of typhoid, and spent much of the latter part of the war as an intelligence officer.[24] He could be very critical of the mentality of the career officers, but was deeply impressed by the courage of the common soldiers. Though he had some time for reading and even reviewed work for the *Revue Historique*, he found it difficult to concentrate as life revolved between a multitude of bureaucratic responsibilities and military action. He served with distinction, receiving Chevalier of the Légion d'Honneur and the Croix de Guerre, and by the end of the war had been promoted to captain.[25]

In January, 1919, Bloch, still in uniform, joined the provisional teaching staff of the repossessed University of Strasbourg along with several other "mobilized teachers," three professors from the German university, and a few professors on temporary transfers from elsewhere. Demobilized in the spring, he officially entered the new Strasbourg faculty as a *chargé de cours* in medieval history in October. Intended as a showpiece of French education and civilization, the university was reopened amidst great pomp with a particularly well-qualified staff. The Faculty of Letters included a number of scholars of great promise as well as some who had already proven themselves. The patriotic mission and challenge of creating a new university helped to make the faculty particularly cohesive and dynamic. It was there that Bloch spent the most productive part of his career, doing the bulk of his writing and being a co-founder and co-editor of the *Annales* with Lucien Febvre.

Like many others of his generation, Bloch was quick to settle into academic life, avoiding direct involvement in the world of politics. At Strasbourg, in particular, many faculty members lived a life apart as they had few ties to the local society and little inclination to develop them. As Bloch later observed of the period following the First World War, "We sold our souls for our rest and our intellectual work with the casualness of men eager to really live after four years of horrors."[26] Bloch quickly formed friendships and working relationships with a number of his colleagues – most notably with the modern historian Lucien Febvre. The seminars of medieval and modern history were next to each other in the Palais de l'Université, and Bloch, Febvre, and their students often went from one to the other. They lived in the same neighborhood and often went home together. Febvre recalled long discussions in the street, often accompanied by another neighbor, the sociologist Maurice Halbwachs.[27]

According to his students, Bloch was very reserved, a marked contrast to Lucien Febvre who was known for his flamboyant lecture style. Philippe Dollinger, who first studied with Bloch in 1923 recalled, "Bloch, with a halting delivery, appeared rather cold, and even distant; his affirmations were qualified with reservations and hesitations, which were somewhat disconcerting to novices eager for certainties. It was only little by little that one became aware of the high value of his erudition and judgement."[28] Not unlike his father, he could be a caustic critic, which some students found upsetting. In addition to his small stature (he was only 5'5"), students were struck by Bloch's elegant dress and most particularly by his eyes, described as mischievous, inquisitive, ironic, and sharp.[29]

Despite the apparent detachment of Bloch and his colleagues, the nation-building mission of the University of Strasbourg meant that political issues could not be avoided entirely. One of Bloch's first assignments as a "mobilized teacher" was to teach French to the German-speaking students who remained.[30] Later he would help promote assimilation by proposing the establishment of scholarships to send the Alsatian students to the interior of France. In addition, he suggested that new teachers occupy positions elsewhere in France before teaching in Alsace-Lorraine. Bloch was also among the Strasbourg professors who taught at the Centre d'Études Germaniques in Mainz during the occupation of the Rhineland.[31]

A staunch supporter of the Third Republic, Bloch's own research contributed to its nation-building agenda, despite its ostensibly apolitical tone. Although he was a strong advocate of a comparative approach, his work remained focussed on the heart of France, helping to give historical depth to the question of what it meant to be French. Though interested in local history, he argued that it must be tied to general history to be of much interest, i.e. not tied to potentially reactionary local interpretations. As a Dreyfusard intellectual, he argued strongly against ethnic and racial interpretations and attempted an objective examination of religious phenomena. He also criticized writings which exaggerated the role of the state at the expense of questions of nationalism and patriotism.

Though never defining himself as Marxist, Bloch was deeply impressed by Marx. In 1936, he wrote to Febvre that he was considering using Marx to bring some "fresh air into the Sorbonne" and that though he suspected that Marx was a "poor philosopher" and probably also an "unbearable man," Marx was without a doubt a great historian.[32] Bloch would, however, later attack those who attempted to establish "Marxist" schools of thought.[33]

Like many of his peers in the Université, Bloch did not take an active part in the major political events of the 1920s and 1930s. Uneasy with the "diplomacy of Versailles and the Ruhr," Bloch nevertheless adopted a neutral position when France invaded the Ruhr in 1923 to collect reparations. Upset about the rise of right-wing political groups, he signed Alain's March 5th Manifesto to the Workers in 1934, without believing that much good would come from it. Although clearly opposed to fascism, Bloch did not favor what he saw as demagogic appeals to the masses to oppose it.[34] As late as September 1937, he wrote to Febvre of his worries about the state of the Left and about Hitler, Mussolini, and Laval, but simply concluded, "One does not see, from one's little corner, the means of doing anything. . . . Please excuse these dark thoughts. It is better, I think, to work [travailler]."[35]

Bloch had a rich home life and his academic and domestic worlds often overlapped. In July, 1919, shortly after he was demobilized, he married Simonne Vidal, whose father was the Inspecteur-Général des Ponts et Chaussées. As Bloch's eldest son, Etienne, has observed, Simonne was from a prosperous family whose affluence undoubtedly also facilitated his career – in terms of comfort, library resources, rest, and so forth.[36] Marriages between *normaliens* and the daughters of successful engineers and businessmen were not uncommon during the Dreyfusard years, as Giuliana Gemelli has observed. Other examples cited included those of Henri Berr, Emile Durkheim, Lucien Lévy-Bruhl, and Léon Brunschvicg. Marc and Simonne were exceedingly close and in addition to sharing family responsibilities often worked and travelled together. Described as cultivated and discreet, timid and energetic, Simonne was devoted to her husband's work. She served him as a secretary, editor, and critic, reading most of what he wrote, attending his lectures and taking notes (much to his students' discomfort), and helping him with his research. This supportive role taken on by Simonne was also, according to Gemelli, not atypical of academic republican households.[37]

Marc and Simonne had six children between 1920 and 1930, and following the deaths of his father and brother, Marc Bloch took on some responsibility for his mother, two nephews, and sister-in-law as well, being seen now as the head of the extended family.[38] Although Etienne remembers his father "working all the time," Marc Bloch, like his father, supervised his children's school work, helping Etienne, for example, with his Latin translations and history compositions.[39]

Contrary to later depictions of the early "Annales School" as being marginal to the historical establishment in France, Bloch and Febvre were in fact well connected to that establishment, contributing regularly to its key journal, the *Revue Historique*, for example.[40] By 1931, just two years after the launch of the *Annales*, the *Revue Historique* included a listing for the *Annales* in its summary of academic journals. In addition, the particular character and mission of the repossessed University of Strasbourg made it second only to Paris in stature, in a very status-conscious system. They did however face an uphill battle to advance further. As Charles-Olivier Carbonell has pointed out, half of the history professors in the Faculties of Letters in 1900 were less than forty-two years old so that many key positions would be unavailable for some time. This generational blockage together with budgetary constraints led to a relatively closed system until the 1930s.[41]

Even though well connected to the historical establishment, Bloch and Febvre did, in the *Annales*, explicitly cultivate an image as outsiders who challenged the character of history as practiced by that establishment.[42] To that end, they sought cooperation from those outside the discipline including geographers and sociologists, who sought higher profiles themselves. In many ways, their criticisms of that establishment were closely tied to the restrictive structures of the Université, which had so dominated the work of their predecessors. The *histoire historisante* or event-oriented history, which they sought to replace was well suited to the requirements of the centrally organized secondary school teaching, which still dominated the Université. As Seignobos, one of those who defined history for the preceding generation, remarked, "events are good instruments of civic education, more effective instruments than the study of institutions." Many of the established historians spent much of their time actively involved in writing textbooks and general histories, rather than in producing new scholarly research.[43]

Geography, on which Bloch and others drew, was also very tied to the school system, limiting the discipline's prestige and affecting its character.

Many of its leading practitioners, including Vidal, wrote school texts. Nevertheless, in pursuit of an independent status, geography was quicker than history to change its orientation, as demonstrated, for example, by the character of the questions asked for *agrégation*.[44]

In contrast to history and geography, which were so defined by the educational structures of the times, sociology was on the margins of the educational system. It was divorced from primary and secondary education, lacking both courses and an *agrégation*. Rather than push for a separate *agrégation*, Durkheim and many of his followers took the *agrégation* in philosophy and tried to make it more supportive of sociology.[45] Even sociology chairs in education were lacking; Durkheim's chair at the Sorbonne, for example, was not named "sociology" until 1913. As Howard Andrews argued, the institutionalization of sociology in France was characterized by top-down processes, diffusing down from the post-secondary level, as opposed to the bottom-up processes that characterized geography and history which not only gave them a strong base but also tied them to scholastic pursuits.[46] Bloch and others connected to the *Annales* were able to profit from sociology's marginal position. They drew on its adherents and also incorporated some of its critiques and methods into a new historical approach to challenge the scholastic and event-oriented history which was dominant at the time of their training.

Bloch, in particular, cultivated not only the geographers and sociologists, but also historians from other countries and did his best to associate them with his work and with that of the *Annales*. Although the period following the First World War witnessed a resurgence of internationalism, French academics remained on the whole fairly isolated.[47] By contrast, Bloch not only travelled widely, but also was an active participant in international conferences including the International Congresses of Historical Sciences in Brussels in 1923 and at Oslo in 1928, and the Semaine d'histoire de droit de Madrid in 1932. In addition, he gave numerous guest lectures abroad at such places as the London School of Economics, Cambridge University, the Institute for Comparative Research in Human Culture in Oslo, and in Belgium at Ghent and at the Institut des Hautes Etudes in Brussels. This contrasted with both the reigning historians and his contemporaries who in many cases remained "hexagonal" (or tied to the "hexagon" of France).[48]

Although the University of Strasbourg was a very stimulating place to be in the 1920s, by the early 1930s Bloch was keen to make the move back to Paris. As the national commitment to the university's special status declined and the political situation became more and more threatening, many of Bloch's closest friends and colleagues contemplated the same move. As a provisional measure to get a foothold in Paris, both Bloch and Febvre in late 1930 unsuccessfully sought a chair at the Ecole Pratique des Hautes Etudes, a research institution modelled on the German seminar system.[49] In 1933, Lucien Febvre left Strasbourg for the Collège de France, making the editing of the *Annales* more difficult. Bloch was also eager to gain access to Parisian schools for his children, writing to Febvre in April, 1936, "When one sent me here [Strasbourg] in 1919, I was not warned that in this colony French would be taught to my kids by men who did not know French at all!"[50] Yet another motive was to reunite his extended families; his mother, his widowed sister-in-law and her two children, and Simonne's three sisters and brother all lived in the Paris region (Simonne's parents had died in the late 1920s).

Bloch's first choice was to obtain a chair at the Collège de France which would free him from heavy teaching responsibilities and give him control over his research and course design; otherwise he would still be greatly constricted by the system of national examinations. In contrast to the faculties and even the ENS, the Collège de France was not so highly constrained by the structures of the Université. Though publicly funded, it also accepted private donations so that some chairs were privately endowed. Its professors were free to choose the subjects and contents of their public lectures and seminars, classes which were given free of charge to audiences that included students from the faculties, foreign students and scholars, and others. By 1933, Bloch had the additional incentive of working with both Febvre and the sociologist François Simiand, who had recently been appointed to the Collège de France.[51]

Bloch entered several competitions: he withdrew in 1928, again in 1933, and was defeated in 1935 in a year-long campaign following the death of Camille Jullian (a campaign confused by the unexpected death of one of the electors, and budget cuts). Finally, in 1936, he withdrew once again in favor of a more promising campaign for the chair of economic history at the Sorbonne. Bloch was officially appointed to the Sorbonne in July 1936 and moved his household there in August.

When Célestin Boulgé, the director of the ENS, became ill in late 1938, Bloch contemplated the possibility to filling his prestigious position. He argued to Febvre, who was not very enthusiastic about the idea, that there did not seem to be other suitable candidates. Boulgé eventually died in office in January, 1940, once Bloch was already in uniform, and was replaced by Jérome Carcopino, one of Gustave Bloch's prize students.[52]

Due to the war, Bloch's time in Paris was to be very short lived. He was called to active duty during the Czechoslovakian crisis in late September 1938 only to be demobilized in early October. He was recalled in the spring of 1939 for a rebriefing, and then called up a final time in late August 1939, following the non-aggression pact between Germany and the Soviet Union. As a man of 53 years and a father of six, Bloch was exempt from the mobilization, but he was determined to serve.[53] Entering the army once again as a captain assigned to staff duties, he joked that he was the oldest captain in the army.[54]

Dissatisfied with his initial posting in Alsace, he eventually obtained a transfer to Picardy, where he soon replaced the officer in charge of fuel supplies. During these long months Bloch continued to do some editorial work for the

Annales and even some writing, particularly during the long days in Alsace. In May, the First Army to which Bloch was attached advanced into Belgium in an ill-fated campaign. Bloch finally saw active warfare and his position in charge of fuel supplies became both crucial and difficult. He was evacuated from Dunkirk shortly before it fell on June 4th. When France fell in late June, Bloch, like so many others, found the shock exceedingly hard to bear. As he wrote to Febvre from his exile in Gurêt, near the family vacation home at Fougères in the unoccupied zone: "It is useless to comment on the events. They surpass in horror and humiliation all that we could have dreamed in our worst nightmares."[55]

In all, the war was a very disillusioning experience for Bloch and one that took its toll both mentally and physically.[56] Even at the outset, Bloch wrote of being mentally weary, lacking the spirit of a man in his twenties. In addition, he found the deprivations and disruptions of army life increasingly hard to bear. Worrying constantly about his effectiveness, he felt daunted by both the task at hand and the tasks ahead. As he wrote to Febvre in October, 1939, "After this war we will be too old to act, even with our brains."[57]

For the next two years, Bloch taught in the unoccupied zone. In late July, 1940, he wrote to several American historians seeking an appointment, all the while expecting to return to Paris in the fall. As the academic year approached he realized the unlikelihood and extreme difficulty of continuing to teach in Paris and managed to obtain a position teaching at the exiled University of Strasbourg in Clermont-Ferrand. The "Statut des Juifs" required special permission for Jews to teach by January, 1941 and through the help of Jérôme Carcopino, Bloch received one of ten exemptions given to university professors for "exceptional services" to the French state.[58]

Meanwhile, in November, 1940, Bloch had received an offer of a two-year appointment at the New School in New York, starting in January, 1941. Though he managed to obtain a temporary leave from Vichy and visas for himself, Simonne, and their four minor children, his two oldest children and his mother faced long delays for their visas. Unwilling to leave without them, Bloch tried unsuccessfully for several months to obtain the missing visas, to book passage for the entire family, and to arrange for transit visas and also the exit visas required for his two oldest sons by recent legislation. To complicate matters, his mother became ill in the spring, dying on April 27, and his wife developed pleurisy. Told that Simonne needed to move to a milder climate, Bloch managed to obtain a transfer to the University of Montpellier in July, and shortly afterwards the New School was forced to withdraw its offer.[59]

Bloch's welcome at Montpellier was a far cry from that at Clermont-Ferrand. There he faced an antagonistic dean, the conservative church historian, Augustin Fliche, and a prohibition on teaching any public courses. Right-wing students rioted at the university in the spring. The following fall, Bloch was again scheduled to teach at Montpellier, but when Germany invaded the unoccupied zone in November, he fled with his family to Fougères.[60]

In contact with various members of the Resistance since his days in Clermont-Ferrand, Bloch joined it sometime before Easter, 1943, as a member of the organization Franc-Tireur. This group, based in Lyon, published an underground newspaper very much opposed to the Vichy regime. Bloch's initial tasks were simply to deliver messages and newspapers, but in July he was promoted to head up Franc-Tireur's operations in the region, Rhône-Alpes, an area covering ten *départements*. He also represented the organization in the regional directory of the Mouvements Unis de la Résistance, a group which coordinated the activities of Franc-Tireur and the two larger resistance groups of Combat and Libération. In addition to being responsible for inspecting his region and organizing its social services, he soon took on the task of preparing the area for the Allies' landing and helped to organize the Liberation Committee, which was to govern the area. By late 1943, Bloch was active head of the region Rhône-Alpes for the Mouvements Unis de la Résistance and was the most senior member on the regional directory. Bloch's arrest came as part of a round-up of key members of the Resistance in Lyon in March, 1944. As the time for the Allied invasion approached, many of those incarcerated in Lyon were deported and others shot as the German forces prepared for retreat. Bloch's turn did not come until ten days after D-Day when he was one of twenty-eight prisoners shot in a field near Saint-Didier-de-Formans, outside of Lyon.[61]

In addition to his responsibilities as a regional leader of the Resistance, Bloch was very much involved during his last days in planning for a new France after liberation. Recalling Bloch's frame of mind during his fleeting visits, Febvre observed, "I found him always the same: lucid, optimistic, and active. Very preoccupied with the days after liberation, and, in particular with the reform, no the revolution, needed in education."[62] For the *Cahiers Politiques*, a Resistance publication devoted to a discussion of the remaking of France, Bloch directed an investigation of educational reform, contributing to it as an author as well, and reportedly even dreamt of filling the position of Minister of Education after liberation.[63]

Educational reform was a subject that had concerned Bloch for some time. Shortly after his appointment to Strasbourg, he had argued that the official programs for secondary education in history, though improved by offering contemporary history, had too little on medieval history through the seventeenth century. In addition, the programs focussed on societies which could be described as an "evolved European type," as opposed to Asian, African, and even other European ones. It was important, he argued, to give the students some understanding of differences in both space and time in order to open their eyes to the "variety of the world." These other civilizations should not be studied simply in the framework of colonial or diplomatic history but "in themselves and for themselves."[64]

In the mid-1930s, Bloch and Febvre, writing as directors of the *Annales*, focussed their attention on the *agrégation*. In 1934 in a text originally drafted by Bloch, they pointed to the under-representation of economic and religious history and to the over-representation of biographical, diplomatic, military, political and administrative history, subjects which were unfortunately detached from their "substrat social." Three years later, they criticized the form and administration of the examination. In place of the lessons *en dehors du programme* which they argued favored repetitors and as such, incompetence, they proposed a program with room for choice, the reinstatement of *explication des textes,* and a much shorter preparation period. The hope was to limit the restrictions which the *agrégation* had imposed on higher education, to allow both more choice and more time for active research.[65]

In his 1943 article for the *Cahiers Politiques*, Bloch attributed France's collapse to a failure of intelligence and character, a prime cause of which was said to be the failure of French society in shaping its youth. He maintained that an overemphasis on professional and technical training had led to small-mindedness and to leaders both limited in their understanding of human problems and afraid of change. Accordingly, the professional *grandes écoles* should be abolished in favor of a more general university education followed by a short period of professional training. As for the universities, they should be, "reconstituted into real Universities, not divided into rigid Faculties which take themselves as 'patries,' but into flexible groups of disciplines."[66]

Bloch argued that the educational system should give a much larger place to more general culture and "things of the spirit" and should encourage research and thinking rather than just cramming for exams. He felt that Latin and Greek had been overemphasized but should be retained in small doses and that a thorough reading of the classical masterpieces in translation was preferable to a painful translation of short passages. Scientific training should not be as technical, with less emphasis on fact and more on the sciences of direct observation, such as botany. As for history and geography, he demanded a teaching that was conceived in a large spirit and that would give "a true and comprehensive image of the world." French students would gain more from learning about the civilizations of India and China than by reciting events of recent French political history. Writing of matters closer to his own work, experience, and spirit, he argued that, "The distant past inspires the sense and respect for differences between men, at the same time as it refines the sensitivity to the poetry of human destinies."[67]

Bloch's parting vision of post-liberation France reflected both the institutional structures within which he had worked all his life and the type of approach which he had come to favor in his own research. That approach drew heavily, but rather differently, on the three changing disciplines of history, sociology, and geography – disciplines which were in turn shaped by the institutional framework of the educational system in France. It is to these issues that the rest of this book is devoted.

PART I

Sociology, geography, and history during Marc Bloch's years of apprenticeship

2

Marc Bloch's training as a *normalien*

I belonged to a school where the dates of entrance made it easier to identify changes. Early on, I found I was much closer, in many respects, to the classes which preceded me than to those which followed almost immediately. My friends and I placed ourselves at the last point of what one could call, I believe, the generation of the Dreyfus Affair.[1]   Marc Bloch. *Apologie pour l'histoire.*

The school to which Marc Bloch referred was, unquestionably, the Ecole Normale Supérieure. By identifying himself with the preceding classes, he associated himself with the classes which were marked by their Dreyfusard sympathies: critical of the army, often anti-clerical, and worried about anti-Semitism. Furthermore, as Alphonse Aulard noted in 1903, most of the students at the ENS were attracted to a moderate form of socialism associated with Jean Jaurès, a sympathy which Bloch shared.[2] The *normaliens* of Bloch's year and their predecessors were also characterized by the nature of their academic work which was shaped in part by both the institutional structures in which they found themselves and the intellectual atmosphere of the times.

Bloch entered the ENS at the age of eighteen in 1904. That was the same year in which the school became united with the University of Paris and was given a new director, Ernest Lavisse. In his opening address to the assembled teachers and students of the ENS, Lavisse spoke to them of their mission. The "social duty" of the *normalien* as educator, which was his true calling, was no less than "the task of creating one France." Lavisse suggested that as teachers, they were to fulfill the same central role as that held by the church in the Middle Ages. He observed that France was in turmoil and that it was up to the educator to point the way and to bring some unity.[3]

In reality, the ENS did not undergo a profound transformation as a result of the recent administrative reforms, which aimed to transform this elite educational institution into a pedagogical center and a place of residence for a select group of students. The *maîtres de conférence,* the only teaching positions at the ENS, were given posts at the Sorbonne and were to teach there in courses open to all the students of the Faculty. However, many professors continued to teach at the ENS after 1904 – often in small seminars which, though technically open to Sorbonnards, were in fact still directed to an almost exclusively *normalien* audience.[4] Despite the rhetoric of the reforms, the ENS was to remain an active intellectual community that was very much aware of its elite status, and in spite of Lavisse's admonition that the *normaliens* take their teaching responsibilities more seriously, a few, as ever, would continue to embark on careers in journalism, politics, and government.[5]

Perhaps the most significant change that came with the reforms was the establishment of a course on the history of French secondary education that was obligatory for all students preparing for *agrégation* at the University of Paris. The professor for this sole obligatory course was Emile Durkheim. As the spokesman for the new science of sociology, Durkheim symbolized the remarkable changes in intellectual atmosphere that had occurred during the previous decades, which accompanied the somewhat less remarkable administrative changes designed to strengthen the French universities.[6] Although the intellectual tone of the entire Faculty of Letters came to be transformed with the growing attention to things scientific, developments in the three fields of sociology, geography, and history will be the focus here.

Appointed to the Sorbonne as a *chargé de cours* of the "science of education" in 1902, Durkheim had previously taught courses at Bordeaux in social science and pedagogy. Convinced of the power of "scientific" reasoning and of the irrationality of the causal explanations of religion, Durkheim attempted to find the basis of a rationalist secular morality; like Lavisse, he sought a new basis on which to build French society. In his obligatory course, given at the ENS, he advocated a form of secondary education that was scientific and secular as opposed to literary and classicist and by so doing he upset many people.[7] The significance of Durkheim's course goes beyond his specific views on education, however, since it gave him the opportunity to introduce his views of the field of sociology to the Parisian candidates for *agrégation*. According to Durkheim, sociology was to become an umbrella science encompassing within its bounds all of the more specialized social sciences. In the process, these former disciplines would become more rigorous and scientific by following the methodology outlined in his *Règles de la méthode sociologique*. A proper use of the comparative method in studying "social facts" could, he argued, lead to the establishment of sociological laws.

Durkheimian sociology, however, had made only minor inroads into the Parisian academic institutions by the year 1904–5, when Bloch entered the ENS. In the public institutions of higher education in Paris, sociology was represented in name only by the chair of *sociologie et sociographie musulmans* at the Collège de France, a post funded by the governments of Algeria, Tunisia, and Morocco and in no way connected with the Durkheimian school. In addition, there were a variety of other social chairs (for example in

Table 1 *Positions using the label "social" or sociologie" in Paris c. 1904[a]*

University of Paris
  Faculty of Letters:
    Espinas   Histoire de l'économie sociale
  Faculty of Law:
    Gide   Economie sociale comparée

Collège de France:
    Le Chatelier   Sociologie et sociographie musulmanes

Conservatoire des Arts et Métiers:
    Beauregard   Economie sociale
    Mabilleau   Assurances et prévoyances sociales

Ecole Libre des Sciences Politiques:
    Cheysson   Economie sociale
    Halévy   Evolution des doctrines économiques et sociales en Allemagne et en
             Angleterre

Ecole des Ponts et Chaussées:
    Gide   Economie sociale

Notes:
[a] This table excludes the Ecole Russe des Sciences Sociales, because of its specialized character, and Ecole des Hautes Etudes Sociales and Collège Libre des Sciences Sociales, both of which were privately sponsored.
Sources: Université de Paris, *L'Université de Paris et les établissements parisiens d'enseignement supérieur, 1903–1904* (Paris: Université de Paris, 1904), p. 100; Université de Paris, *Livret de l'étudiant, 1905–1906* (Melun: Imprimerie Administrative, 1905), pp. 231–232; Université de Paris, *Livret de l'étudiant, 1904–1905* (Melun: Imprimerie Administrative, 1904); *Minerva: Jahrbuch der Gelehrten Welt*, 1904–1905 (Strasbourg: Karl Trübner, 1905), pp. 837–875.

économie sociale, in *histoire du travail*, and in *assurances et prévoyances sociales*) held not by Durkheimians but rather by men adopting such perspectives as those of Frédéric Le Play, of political economy, and in the case of Alfred Espinas, of a biological interpretation of social phenomena (see Table 1). Privately sponsored ventures included lectures at the Collège Libre des Sciences Sociales and the Ecole des Hautes Etudes Sociales, which received some government funding and drew on a number of university professors including, in the second case at least, some of the Durkheimians.[8] In addition, René Worms sponsored such activities as the Institut International de Sociologie, the *Revue Internationale de Sociologie*, and the Société de Sociologie de Paris, which drew together a great variety of people with some interest in sociology, however defined.[9]

Those Parisian professors most closely associated with the growing

Durkheimian school held a variety of positions, none of which were officially labeled as posts of sociology or even of *économie sociale,* and most of which were fairly minor. Durkheim at this point was still only a *chargé de cours* at the Sorbonne, and Marcel Mauss and Henri Hubert were both *maîtres de conferences* in V[th] section ("religious sciences") of the Ecole Pratique des Hautes Etudes. Among those who were not strictly speaking "Durkheimians" and yet were sympathetic to their views were Elie Halèvy (who taught at the Ecole Libre des Sciences Politiques), Antoine Meillet (a linguistics professor in the first section of the Ecole Pratique des Hautes Etudes), and Lucien Lévy-Bruhl (a *chargé de cours* for Emile Boutroux's chair of the history of modern philosophy at the Faculty of Letters). By 1907–8, Bloch's last year at the ENS, the situation had begun to improve for Durkheim and his associates. At the Sorbonne, Durkheim became a titled professor; Célestin Bouglé took over Espinas' chair; and Lévy-Bruhl obtained the chair of modern philosophy. Also, at the Ecole Pratique des Hautes Etudes, Mauss became a *directeur d'études* followed by Hubert in 1908.[10]

Despite their marginal positions, the Durkheimians had founded two journals which helped both to develop and to diffuse their ideas as well as to give them some identity and legitimacy as a school. The first and most influential was the *Année Sociologique.* Directed by Durkheim, its first series ran from 1898 to 1913. The journal contained both lengthy *mémoires originaux* and numerous "analyses" and "notices" of literature published each year in a great variety of fields thought to be of interest to the growing field of sociology. By contrast, *Notes Critiques: Sciences Sociales,* directed for its first six years by François Simiand, was restricted to relatively short reviews, along with bibliographical notes and a few editorial comments. Drawing on many of the same collaborators as the *Année Sociologique,* this journal was supposed to focus more directly on the social and political issues of the day – though it only did so to a limited degree.[11] It also appeared much more frequently (eventually ten times per year) while the *Année Sociologique* appeared only annually. The impact of these journals was considerable, particularly on the young *normaliens* of this period. Marc Bloch, for example, would later testify, "To the old *Année* the historians of my generation owed more than they knew how to say," and would identify himself with "all those who found in the *Année Sociologique* of yesteryear one of the best intellectual elements of their years of apprenticeship."[12]

Although sociology remained marginal at the University of Paris in 1904, geography held a more favorable position even though it continued to be institutionally tied to the study of history.[13] Before the expansion of the faculties in the late nineteenth century, geography was represented at the University of Paris only by Auguste Himly's chair, but following the appointments resulting from the union of the ENS and the Sorbonne, positions in geography included two full professors in the Faculty of Letters (Vidal de la

Blache and Marcel Dubois), one in the Faculty of Sciences (Charles Vélain) as well as two *chargés de cours* (Lucien Gallois and Augustin Bernard), and a *maître de conférence* (Henri Schirmer) all in the Faculty of Letters (see Table 2). Other positions could be found at the Collège de France, the Ecole Pratique des Hautes Etudes, the Ecole Libre des Sciences Politiques, and the Ecole des Mines. During Bloch's tenure at the ENS, the positions in geography in Paris remained remarkably stable. By the academic year 1907–8, the only changes were minor shifts all outside of the University of Paris.[14] Nevertheless, geography maintained a much stronger institutional base than sociology.

Himly, in particular, had played a very important role in late nineteenth century geography in Paris. As a practitioner of *la géographie historique,* he was convinced that geography's proper role was in the service of history. According to Vincent Berdoulay, he was one of the opponents of any proposal for a separate *agrégation* in geography, an opposition which meant that the common *agrégation* for history and geography remained. Suspicious of attempts to make the field more scientific, he helped to exclude physical geography from the Faculty of Letters where he was dean from 1881 until his retirement seven years later; when a course in physical geography was introduced in 1886, it was placed in the Faculty of Science.[15]

By 1904, however, the tone of geography at the University of Paris had changed significantly. On Himly's retirement in 1898, Vidal de la Blache was appointed to the Sorbonne and in turn Vidal's geography position at the ENS was filled by one of his former students, Lucien Gallois. In contrast to Himly and despite his own training in classical history, Vidal became attracted to the physical sciences and did much both to make geography appear more "scientific" and to separate it from history. In so doing, Vidal profited from a variety of institutional shifts in the late nineteenth century French education system.

In the aftermath of the Franco-Prussian war, a major priority of the nascent administration for the Third Republic was to reform the national education system. Jules Simon, then Minister of Education, led the reform efforts. Geography was given a central role because of its perceived importance for strengthening the nation. In addition, following the loss of Alsace-Lorraine, a new chair, transferred from Strasbourg, was created at Nancy and there was considerable pressure to make that chair a specifically geographical one, as opposed to the combined history and geography chairs that were then the norm. That chair was filled by Vidal in 1872[16] shortly after his successful doctoral defence. Only five years later, he replaced Ernest Desjardins at the ENS. One of Vidal's teachers, Desjardins, who was a classicist and archeologist, had taught geography at the ENS since 1862 and was very much within the historical geography tradition. The pressures, political, ideological, and institutional, were clearly there for a strengthened and distinct field of geography. Vidal responded, leading him rapidly to the head of a revised field.[17]

Given his classical training, Vidal's initial contributions did not depart

Table 2 *History and geography at the University of Paris, 1904–1905*

Positions in History at the Faculty of Letters

professeur

Bouché-Leclercq	Histoire ancienne
Guiraud[a]	Histoire grecque
Bloch[a]	Histoire romaine
Luchaire	Histoire du Moyen Age
Monod[a]	Histoire de la civilisation et des institutions du Moyen Age (suppléance Pfister)
Lavisse	Histoire moderne (suppléance Seignobos)
Bourgeois[a]	Histoire politique et diplomatique des temps modernes
Rambaud	Histoire moderne et contemporaine
Aulard	Histoire de la Révolution française

professeur adjoint

Denis	Histoire contemporaine
Langlois	Sciences auxiliaires de l'histoire

chargé de cours

Pfister[a]	Histoire de la civilisation et des institutions du Moyen Age (chair: Monod)
Seignobos	Histoire moderne (chair: Lavisse)
Grébaut	Histoire ancienne des peuples de l'Orient
Revon	Histoire de la civilisation et des peuples de l'Extrême Orient
Diehl	Histoire byzantine

Positions in Geography at the Faculties of Letters and Sciences

professeur

Vidal de la Blache	Géographie
Dubois	Géographie coloniale
Vélain	Géographie physique (Faculty of Sciences)

chargé de cours

Gallois[a]	Géographie
Bernard	Géographie et colonisation des peuples de l'Afrique du Nord

maître de conférence

Schirmer	Géographie

Notes:
[a] Associated with the ENS prior to merger.
Sources: (in order of usefulness and accuracy): Académie de Paris, *Rapport sur la situation de l'enseignement supérieur & Rapports sur les travaux et les actes des établissements d'enseignement supérieur, 1903–1904*, pp. x–xii, 184–191; Académie de Paris, *Rapport sur la situation de l'enseignement supérieur, 1904–1905*, pp. xv–xvi; Université de Paris, *Livret de l'étudiant*, 1904–1905, 1905–1906; *Minerva*, 1904–1905.

greatly from the then reigning historical geography, but once at Paris, he would help point the way toward a more scientific view of the discipline. In this endeavor, the *Annales de Géographie*, founded by Vidal and Dubois in 1891, was central. The early issues were dominated by articles on physical geography with secondary emphases on exploration and colonial geography, subjects of particular interest to Dubois.[18] The colonial theme had characterized earlier journals such as the *Explorateur* (1875), the *Revue Géographique Internationale* (1876), and the more prestigious *Revue de Géographie* (1877), but the emphasis on physical science was new. As laid out in *l'avis au lecteur* of the first issue, the journal aimed, among other tasks, to familiarize geographers with the relevant findings of such fields as geology, meteorology, and natural history. The journal was to be a "scientific" one, with a goal to professionalize the discipline of geography and to take it out of the hands of both the local geographical societies and wealthy explorers.[19]

To many of the ENS students, Vidal's teaching appeared as a refreshing change. As Lucien Febvre recalled, Marc Bloch, like many of his contemporaries and predecessors, was quite taken by Vidal's teaching. Describing the attraction which this new geography held, Febvre wrote:

Geography was many things without a doubt, but for the young Frenchmen shut in classrooms, in ugly and sullen study rooms (the bottom painted chestnut, the top in dirty ochre, and above the bent heads, the pale and stifling light of gas) . . . geography was fresh air, a stroll in the countryside, the journey back with an armful of broom and foxglove, eyes cleaned out, brains washed, and the taste of the "real" biting the "abstract."[20]

Colonial geography as represented by Dubois and his students provided an alternative interpretation of the field but could not compete in prestige or following. Also striving to become more "scientific," Dubois' approach was that of an applied social and economic science but without the same emphasis that Vidal gave to physical geography. In the other Parisian institutions, historical, colonial, and commercial geography reigned.

By the time of Bloch's entry to the ENS, history had expanded dramatically at the university, changing in both content and approach. Before 1878, the Sorbonne had only two chairs of history, one for ancient history and one for modern history. After the 1904 promotions associated with the union of the ENS and the university, positions in history at the Sorbonne included three for ancient history, two for medieval, five for modern and contemporary, three for Byzantine and Oriental history and one for "the auxiliary sciences of history" (see Table 2).[21] In addition, there were positions in the history of art and in archeology at the Sorbonne. Elsewhere in Paris, there were numerous other history positions at such institutions as the Ecole des Hautes Etudes, Collège de France, Ecole Libre des Sciences Politiques, and Ecole des Chartes. By 1907–8 history had made further gains at all levels including several new positions in religious history. These new posts helped to consolidate history's very strong position and clearly surpassed the much more modest institutional gains made by sociology and geography.[22]

In part this success reflected the very active role which historians had played in the reform movements. In addition, it reflected the connection established between history and the molding of a secular national consciousness. Even Durkheim suggested a central role for historical studies in the secondary schools until studies in sociology were more advanced. Given the close institutional ties between secondary and higher education in France, a stress on historical studies at the secondary level led to more course offerings and chairs at the higher level since a major function of the Faculties of Letters was to train teachers.[23]

This blossoming of history at the Sorbonne was accompanied by a change in approach and methodology. During the course of the nineteeenth century, an impressive growth in engineering and technology and the rising international reputations of such French scientists as Edmond and Henri Becquerel, Claude Bernard, Marcelin Berthelot, Jean Martin Charcot, Louis Pasteur, and Pierre and Marie Curie, helped to create an atmosphere where science was revered. The historians, however, lacked the prestige of their scientific colleagues. Increasingly historians became convinced that their profession would advance only by adopting a more rigorous methodology, a methodology which often, not surprisingly, took on scientific trappings.[24]

This movement to establish history on a more scientific basis was not entirely new; what was new was the opening of the Sorbonne to such ways of thinking.[25] The growing acceptance of the new approach was marked by the establishment of the first historiography course in 1896–7, taught jointly by Charles Seignobos and Charles-Victor Langlois. In 1898, they published what was to become the basic handbook for historical training at the Sorbonne, *Introduction aux études historiques*. Contrasting history to the sciences based on direct observation, this manual laid out the principles of the historical or "indirect" method which, they claimed could also lead to "scientific knowledge." Essentially that method was one of documentary criticism based on the careful examination of primary sources to reconstruct the past. Methods of "external criticism" could be used to establish the source, authenticity, and authorship of the document. Once the origin of the document was established, methods of "internal criticism" could be used to establish the probability that the statements in the text were true. To do this, first one must determine what the author meant and then attempt to verify the author's statements by questioning their accuracy, and searching for contradictions both within the document and between it and other documents based on independent observations. Following the determination of the facts by these methods of external and internal criticism, the historian could attempt to "organize them into a body of science" by a series of "synthetic operations."[26]

Although this key course and manual were jointly done, the two historians differed somewhat in their contributions. Seignobos, a former *normalien* and in part a product of its tradition of literary scholarship, promoted what he saw as a particularly French form of historical science – that which did not neglect the "scientific synthesis" so often overlooked by German scholars who had done so much to develop the methods of documentary criticism. Nevertheless, like those in Germany whom he criticized, Seignobos advocated a search for "uncontestable formulations" based on "a quantity of detailed facts." In this way, Seignobos hoped to avoid the "puerile generalizations" associated with earlier French work.[27]

Langlois, on the other hand, had been trained at the Ecole des Chartes and was far less critical of the German tradition; his contribution was to help to transfer the methods of the "chartiste" to the Sorbonne, putting more emphasis on an examination of sources than on synthesis. Labeling the new spirit of the Sorbonne as a more "scientific" one, he praised the concept of the "Faculty as a workshop" seeking to teach the scientific methods in new historical laboratories. Langlois was among those who supported the establishment of the *diplôme d'études supérieures*. This new degree, which was made a prerequisite for the history *agrégation* in 1894, was designed to demonstrate research ability, and required a written *mémoire*.[28]

Other indications of the attempts to establish history on a firmer basis can be found in the creation of journals and professional societies. An early example is the *Revue Critique d'Histoire et de Littérature* begun in 1866, a journal composed of critical reviews which attempted to set a scholarly standard. A more ambitious undertaking was the *Revue Historique* (1878), which published scholarly articles as well. Its aim was to promote "objective history" in contrast to the Catholic *Revue des Questions Historiques* (1866) which was seen as marred by religious and political biases. The success of the *Revue Historique* and the progress in documentary investigation encouraged the founding of a number of more specialized journals.[29]

Of the turn-of-the-century journals and professional societies, the *Revue d'Histoire Moderne et Contemporaine* (1899) and the Société d'Histoire Moderne were particularly ambitious and became important for the debates over the nature of historical inquiry. Pierre Caron and Philippe Sagnac were instrumental in both ventures. They hoped to encourage an application of the new "scientific" methods to what they saw as the neglected domains of modern and contemporary history and by so doing sought to exclude the "counter-revolutionary passions" which had marred their field. One object of such an attack was the Société d'Histoire Contemporaine, an organization of non-academic historians founded in 1890. In contrast to the "pure journalism . . . without scientific value" represented by that society, the new Société d'Histoire Moderne was to exclude "simple amateurs" allowing only "those who apply the rational method in their work" and excluding preconceived arguments in support of particular political or religious doctrines.[30] In spite of the rhetoric and the greater attention to method, the Société d'Histoire Moderne was marked with a Dreyfusard stamp and a clear commitment to the principles of the Third Republic.

By the time Marc Bloch entered the ENS, history was very well established in the university system, competing favourably with letters and surpassing philosophy in the number of chairs. It had also changed considerably, assuming scientific airs and becoming more scholarly and specialized, with an increasing emphasis on the modern and contemporary periods. Bloch's class was not immune from this atmosphere. The majority of *mémoires* submitted for the *"diplôme d'études supérieures"* (DES) in history and geography at the University of Paris in 1906 (a year including many of his classmates) looked at the eighteenth and nineteenth centuries. Although political and diplomatic history was still very common, a few had begun to choose topics in social and economic history.[31] Bloch's *mémoire*, completed the next year following his leave for military service, was on the possessions of the Chapter of Notre Dame in two archdioceses to the south of Paris in the thirteenth century. There he addressed questions of economic and social life, examining the decline of serfdom, migration between domains, and the role of capital in the economic life of these rural areas.[32]

These topics contrasted with those chosen just nine years earlier, shortly after the DES became obligatory for candidates for *agrégation* in history and geography. Then, the medieval period and the seventeenth century predominated, with popular topics being the leading figures of the Middle Ages and the dealings of Colbert and Mazarin.[33] Comparable information is obviously lacking for the years before the DES was instituted but it is interesting to note that of the historical studies published by university historians between 1866 and 1875, over half treated ancient history, just over a quarter medieval, and the rest concerned the periods of the sixteenth to the nineteenth centuries.[34]

The interests declared by Bloch's colleagues in the 1904 ENS *promotion* also demonstrated both their attraction to the new trends in history and the drawing powers of the less-established fields of sociology and geography. Of Bloch's six colleagues in history and geography, Georges Hardy was listed in the ENS records as specializing in modern history and Nicholas Neirs in contemporary history. François La Vieille was interested in economic and social history, Laurent-Vibert in ancient history, Philippe Arbos in geography, and Marcel Granet planned to do some studies in law. Out of a class of twenty in letters, two (Jean Ray and Maxime David) contributed to the first series of the *Année Sociologique*, and both Granet and Bloch were also strongly influenced by the developing field of sociology. This attraction to Durkheimian sociology reflected the interest which Durkheim generated among the students shortly after his arrival in Paris in 1902.[35] By contrast, the number during

Bloch's period who would follow careers in geography was relatively low. Whereas during the years 1876–95 there had been usually at least one or two ENS students per *promotion* who would later become distinguished as geographers, for the entire period from 1901 to 1907 there was only one – Philippe Arbos.[36]

Despite their fascination with the new methods and topics, Bloch and his colleagues could not ignore the requirements of *agrégation* if they hoped to gain access to the better teaching posts. Although the examination topics did change to reflect new interests, such changes were slow in coming. For Bloch's year the jury, led by Langlois, stressed political history but indicated that other questions might be mentioned – as in a sketching of the political and social situation of the Roman world for a question on Caius Gracchus and of the economic and artistic consequences of the fourth Crusade. On the other hand, for the geography question, the Mediterranean region of France, candidates were expected to demonstrate a thorough grounding in the relatively new fields of physical and human geography, discussing the economic resources of the region and the conditions of settlement (*peuplement et habitation*). Similarly for the oral exam, the history questions stressed political and diplomatic history, only touching on social history (particularly for ancient history), and the geography questions were more closely related to the intellectual trends of the times focussing on physical and regional geography rather than historical geography.[37]

To get a clearer picture of the situation in which Bloch and his colleagues found themselves, one must look beyond the institutional status of history, geography, and sociology in Paris. Among the other developments that colored the training and experiences of the young *normaliens* were the push for greater interdisciplinary communication, the quality of the provincial universities, and their changing relationships with the *sociétés savantes* (learned societies).

Greater interdisciplinary communication had been a high priority of many of the university reformers. Louis Liard, for example, had often argued that one reason to reunite the faculties into a university body was to encourage communication between them and in so doing to provide a fruitful atmosphere for scientific discoveries, which were often made at "the borders of the sciences." By 1909, Liard noted that some progress had been made: through the influences of Bernard and Pasteur, the Faculty of Medicine had become more of an experimental science; the Faculty of Law had become more open to the historical method; and students no longer limited their training to one faculty. However, much remained to be done. Conceding that the various disciplines would maintain their particular characters, Liard continued, nevertheless, to argue for an opening up of the barriers between them and to express hope for future cooperation.[38]

This theme was taken up by other reformers including not only Lavisse, but also Ferdinand Lot, a history professor at the Ecole Pratique des Hautes Etudes and a man with whom Marc Bloch would remain in contact for years.[39] Stressing the institutional barriers to interdisciplinary exchange, Lot attacked the faculty system which included such absurd divisions as those within the field of geography, split between the Faculties of Letters and Science, and within sociology, split between Law and Letters. To rectify this situation, he suggested the creation of institutes, which could group together related studies, and even a Faculty of Sociology combining the Faculties of Letters and Law. Arguing that students suffered by specializing too early, Lot also proposed a much wider adoption of a system similar to the PCN in sciences, which required medical students to have a thorough grounding in physics, chemistry, and natural sciences.[40] Although radical institutional changes were not made, many shared this belief in the value of interdisciplinary exchange.

Another reaction to the increasing compartmentalization of knowledge was the creation of a new journal in 1900, the *Revue de Synthèse Historique*, which was founded by Henri Berr, a professor of rhetoric at Lycée Henri IV. Berr hoped to draw together "the diverse historical sciences" seeing history, not sociology, as the leader of the "human sciences." To justify this position, he argued that all social science should draw on the "concrete facts of history" and that there was much of interest in history that was not included in sociology. Nevertheless, Berr also expressed an interest in discussions of theory and most of all in "synthesis," which he saw as the highest task of historical science.[41] Articles published in the *Revue de Synthèse Historique* included lengthy discussions of the scientific status of history as well as discussions of the methodology in such fields as sociology and geography.

This new journal was greeted with enthusiasm by history students, as it represented a welcome change from the tedium of the documentary criticism to which they were commonly subjected. Lucien Febvre has described his experience on discovering the journal at the ENS library in 1902:

This review which proclaimed that another history than the history of battles, diplomatic treaties and political intrigues existed and ought to exist – this review which proclaimed and realized a design to gather together historians and archivists, geographers and ethnographers, linguists, economists, and philosophers for a work of synthesis, all fraternally united in the concern of the common work – this review which installed enthusiasm and hope where others knew only how to distill boredom: dear friend, dear friend, what liberation and joy![42]

In addition to pushing for greater interdisciplinary communication, the university reformers, with their desire for nation building, also strove to strengthen the provincial universities. Those universities continued to lag far behind both the University of Paris and their German counterparts. Liard, for example, wrote in 1890, "In our time once the ramparts of Strasbourg were enlarged, Germany did not have any concern more pressing than to build behind them a vast University, like an advance fortress against the spirit of

France. She knew by experience what Universities could do to the mind and how they contribute to forming the soul of nations."[43] To facilitate this task, Liard advocated stronger links between the provincial universities and their milieux and noted that the universities could help solve regional problems by developing practical applications of their science.

Liard and others had hoped that by establishing as few as six or seven universities in the provinces they could be made strong enough to accomplish all that was expected of them. However, local interests in those towns which were unlikely to be so designated prevented such a measure, and in the end fifteen universities were sanctioned when the universities regained corporate identity in 1896 following a century of being disabled.[44] Despite somewhat increased funding at both the national and local levels, these new universities failed to meet the aims of either the ministry or the regions. In a 1905–1906 report, for example, Lot complained of poor facilities, noting the contrast between "the palace of granite and of marble" of the German Faculty of Letters at Strasbourg and "the stable which bears this name at Nancy." He recorded glaring inadequacies in staff and in courses on such important subjects as history and geography, and argued that to combat the influence of the clergy, each provincial university needed at least one professor of "religious sociology." Representatives of the regionalist movement had their own complaints; the provincial universities, they charged, remained unable to compete with the Sorbonne and fell far short of their appropriate role as "the living soul" of a revitalized region.[45]

Reverberations of the struggle to upgrade the provincial universities were also felt on another closely related front, that of the *sociétés savantes*. Initially a bastion of the aristocracy, these societies began to admit representatives of the local bourgeoisie and some faculty members. The more ambitious professors, however, ignored the societies and denigrated their dilettantism. In the mid-nineteenth century, the Ministry of Public Instruction, seeing the societies as a threat, took over their direction through the Comité des Travaux Historiques. This committee issued "instructions" enumerating topics deemed worthy of investigation, awarded grants, established a journal (the *Revue des Sociétés Savantes*), and instituted a Congrès des Sociétés Savantes to be held annually at the Sorbonne.[46] By the 1890s as the strengthened faculties came to depend more directly on local financial support, their professors were encouraged to partake more actively in local affairs including those of the *sociétés savantes*. Attention was to be paid to subjects such as local history that had long been abandoned to the societies. Courses in regional history began to be established (see Table 3). By taking the initiative, university professors were not only to appease local interests but also to provide guidance in the new scientific methods and to link local history to the broader questions of interest to the history of France as a whole.[47]

At the turn of the century, some Dreyfusard historians such as Maurice

Table 3 *Positions of local history in the French Faculties of Letters,
1891–1904*

Faculty of Letters	Title of post	Holder of post, rank and date
Positions listed in 1891[a]		
Bordeaux	History of Bordeaux & Southwest France	Jullian, cc 1891, prof 1897
Clermont-Ferrand	History of Auvergne	Rouchon, mc 1891
Poitiers	History of Poitou	Richard, mc 1891; Boissonnade, prof adj 1897
Rennes	History of Brittany	de la Borderie (cours libre)
New positions listed between 1892 and 1904		
Toulouse	History of Southern France	Molinier, prof 1893
Nancy	History of Eastern France	Pfister, prof 1894
Aix-Marseille	History of Provence	Clerc, prof 1895
Caen	History of Norman Literature & Art	Souriau, prof 1896
Dijon	History of Burgundy	Kleinclausz, cc 1897
Marseille	History of Provence	Clerc (see Aix), cours annex, 1898
Lyon	History of Lyon	Charléty, mc 1899
Besançon	History of Franche-Comté to 1790	Gauthier, cours libre, 1900
Lille	History of Northern France	de S. Léger, mc 1901

No positions to 1904: Montpellier, Grenoble, Paris.

Notes:
[a] = 1st volume of *Minerva*; cc = chargé de cours; mc = maître de conférence; prof =
professeur; prof adj = professeur adjoint.
Source: Minerva: Jahrbuch der Gelehrten Welt, 1891–1904.

Dumoulin and Aulard argued that even greater control was needed over the
sociétés savantes, which they saw as politically conservative and scientifically
backward. As a solution Dumoulin proposed a federation of the societies to
introduce, as he put it, some centralization into the decentralization.
Speaking before the Congrès des Sociétés Savantes, Aulard charged that the
societies had not yet learned to choose interesting and important topics and
reminded his audience, "You are modern Frenchmen, sons of the Revolution
of 1789 and servants to its principles."[48] Soon afterwards, the newly formed
Société d'Histoire Moderne (of which Aulard was a founding member)
began an investigation into ways to direct work in local history. Proposed
centers for that direction included alternatively the Comité des Travaux
Historiques (which could use sanctions to enforce its "instructions"), the

universities, the various departmental archives, and the Société d'Histoire Moderne itself.[49]

What then did these struggles over the provincial universities and over the control of the *sociétés savantes* mean for a young *normalien* in history? On the one hand, the massive discrepancies between the resources of Paris and those of the provinces meant that the Parisian student had a decisive advantage over his provincial counterparts in the competition for key lycée positions in the educational hierarchy (particularly if he came from the ENS).[50] On the other hand, even a *normalien* could expect to spend some, if not most of his career teaching in the provinces. Describing this disillusioning experience, Dumoulin wrote, "I know from experience these hours of discouragement and dreariness, which cast a shadow over the first moments of the provincial life of the young professor. Up to now one had been in a hothouse: Ecole Normale, Sorbonne or Faculty . . . Suddenly one falls into a peaceful milieu that is anything but scholarly. . ."[51]

Urging the young professors to interest themselves in the history of the region in which they were located, Dumoulin suggested that they could do much with the limited resources at hand and suggested ways of using the *sociétés savantes* and the local archives, libraries and museums. Provincial history needed to be rewritten, and the young *universitaires* had the necessary training to link it to "general history." One such effort to rewrite French history was a series of lengthy articles published in the *Revue de Synthèse Historique*, "Les Régions de la France," which drew not only on the newly established professors of regional history, but also on two young *normaliens*, Lucien Febvre and Marc Bloch.

In the aftermath of the Dreyfus Affair, as in the years following the Franco-Prussian War, high hopes were placed in the academics. They were both to strengthen national consciousness and prove to the world that France was still a leader in the realm of thought and learning. In an effort to compete with foreign scholars, to match the glowing reputations associated with the likes of Pasteur and the Curies, and to roust out clerical and reactionary interpretations, the disciplines in the Faculty of Letters were to become more scientific. Practitioners in the fields of sociology, geography, and history responded to these pressures in differing ways – the first striving for "scientific laws" through the comparative method, the second using the data of the physical and natural sciences, and the third stressing "exactitude," documentary criticism and occasionally "scientific synthesis." With the growth of the academic professions and the struggle to become scientific, some fields became very specialized, prompting warnings that the most significant scientific discoveries came not from highly specialized research but instead from interdisciplinary communication.

Within the field of history, the task of nation building had some specific implications that went beyond the adaptation of a new scientific approach and an integration of findings from other disciplines. The topics of research seen as most important began to change. Rather than studying ancient history, some young scholars were now to turn to a reinterpretation of modern and contemporary history, whose controversies lay at the root of much of the unrest in France. Furthermore, local and provincial history, so long regarded as trivial, became prime targets. One could no longer leave it to the provincial *savant*, who might well use it to cultivate reactionary views, but instead one must draw closer links between that history and the history of France as a nation, using the methods of historical science to give backing and prestige to the new interpretations.

The generation of 1905

To return to Bloch's opening testimony, his identification with the generation of the Dreyfus Affair was in part a disassociation with the classes "which followed almost immediately." In contrast to that of the late 1890s, the "generation of 1905," as it has been labeled,[52] was marked by a growing dissatisfaction with the scientism, anti-clericalism, and socialism of their predecessors, accompanied by a renewed interest in the irrational, the spiritual, and Catholicism. The year 1905 was significant in part because of the growing political unrest on the international scene. During the first Moroccan crisis, Germany once again demonstrated a military challenge to France, and the Russian Revolution of 1905 left many with a sense of unease. Within France, the official separation of church and state in 1905, along with the use of spies in the army to check for dangerous clerical tendencies, led some to question their Dreyfusard leaders.

In Parisian circles one important intellectual leader of the new generation was Henri Bergson, who was appointed to a chair at the Collège de France in 1904 (following two unsuccessful bids for a chair at the Sorbonne). His course soon became extremely popular. He drew full-house attendance and was seen by many as a refreshing change from the "scientific laboratories" which had come to dominate teaching in the social sciences and even some of the humanities at the University of Paris. In contrast to Durkheim's science of society, Bergson spoke of "personality," "intuition," and *élan vital.* Among those attending was Henri Massis, who had enrolled in the Sorbonne in 1904 but, in contrast to Bloch, rejected much of what he found there and soon became one of the leading critics of the Nouvelle Sorbonne.[53]

Massis had begun his initiation into the field of literary "science" with a study of the "method of work" of Zola based on documents recently placed in the Bibliothèque Nationale. According to Massis, this experience was designed not to form his taste nor to cultivate his spirit, but instead to recruit "workers of science." Subsequently, when Gustave Lanson proposed that he prepare a paper dating ten of Voltaire's letters, "he fled, disgusted."[54] He began to draw his inspiration from scholars outside the Sorbonne, turning not only to Bergson, but also to Anatole France, Maurice Barrès, Paul Claudel, and Charles Péguy.

In 1910, after publishing an article attacking the Germanization of French culture, Massis joined forces with Alfred de Tarde, a law student whose father, Gabriel Tarde, had been engaged in a long debate with Durkheim. Together, under the pseudonym of "Agathon," Massis and Tarde wrote a damning series of articles attacking the premises of the Nouvelle Sorbonne, including what they saw as its scientism, obsession with German thought, and belittlement of the classics in a misguided attempt to do away with an academic elite. Seeking to restore the humanities to their proper character, they ridiculed the language of the leaders of the Nouvelle Sorbonne, who sought to reduce scholars to manual laborers practicing their trades in "workshops" and "laboratories." Among the objects of attack were those professors who had led the struggle to make their fields more scientific, including Durkheim, Langlois, Seignobos, and Lanson. To demonstrate the harm that had been caused, Massis and Tarde pointed to the increasing testimonies of the poor quality of writing and speaking in examinations.[55]

Much to their surprise, their articles provoked public answers from the likes of Lavisse, Aulard, and Alfred Croiset (the dean of the Paris Faculty of Letters), and soon much of Paris was buzzing with talk of "the crisis of French" and of the possible identity of "the mysterious Agathon."[56] Even more surprising to Tarde and Massis was that the spokesmen for the Nouvelle Sorbonne began to admit that there might be a problem. Within the Sorbonne support came mostly from literary and law students but drew students from other fields as well. For example, a philosophy student wrote to *L'Action*, "Certainly, the Sorbonne professors no longer consider the intimate life of the masterpieces. They wanted to reduce history, literature, and philosophy to I don't know what order of knowledge both dry and dead."[57] Further support came from another disillusioned student, René Benjamin, who soon published a scathing satire of the Sorbonne, *La Farce de la Sorbonne*, singling out Aulard, Seignobos, and Victor Basch, a professor of "aesthetics and the science of art." Pierre Lasserre, a literary critic associated with the right-wing Action Française, devoted less attention to an attack on scientism and more to the anti-clericalism which he felt underlay the harmful changes which characterized the Nouvelle Sorbonne.[58] Like critics within the system such as Liard and Lot, these new critics also attacked the over-specialization of the Sorbonne and argued for a synthetic approach, based, however, not on science but instead on the values and insights of literary classics.

Even the ENS began to show some signs of the changing atmosphere despite the continued adherence of its teachers and administration to the principles of the Nouvelle Sorbonne. "Agathon" reported that in 1912 approximately forty of the *normaliens*, constituting a third of the student body, were

practicing Catholics, a marked contrast to the situation eight to ten years previously, when only three to four were to be found. Although demonstrating a
shift to the right, the *normaliens* as a group did not go so far as to identify with
the Action Française, which counted only two *normalien* members in 1912.[59]
Many were still interested in socialism, albeit at times a Christian socialism
rather than that of Jaurès.

Marc Bloch, who identified so closely with his predecessors, was not left
untouched by this reaction. Though adopting many of the methods and aims
and much of the terminology of his scientist teachers, he too would come to
question the applicability of some of their views to history and even more
broadly to *les études humaines.*

3

History under attack

In February, 1906, while still a student at the ENS, Marc Bloch began a short notebook entitled "Historical Methodology." He began with the words, "I would like to fix in writing certain ideas on historical methodology which have developed in my mind during the last while, but which still take on a vacillating form and vague contours."[1] And indeed the positions which he laid out were somewhat ambiguous and unclear. Given the swirling debates that had developed within history and between history and sociology, this is not surprising. Since the methodological issues involved in these debates would resonate throughout Bloch's later works, it is important that their genesis and the key actors are understood before turning to his notebook and to the many writings which followed.

The French historical establishment during Bloch's years of training was characterized by its focus on political history. In part this was a consequence of the attempt by the historical profession to become more scientific, since political topics were more amenable to the type of documentary criticism of archival sources which was now required as a mark of one's professional training. In addition, political history was favored by the increasingly close ties between the reigning historians and secondary education. Many of these historians devoted much of their time to the writing of school manuals and to the preparation of students for *agrégation*, at a time when political history was seen as central to the creation of a unified France under the auspices of the Third Republic.

The dominance of political history was not, however, without its challengers from both within and without the profession. Two of the leading nineteenth-century historians, Jules Michelet and Numa Denis Fustel de Coulanges had written social histories which were still praised by the turn-of-the-century historians. Among their admirers were Gabriel Monod, a co-founder of the *Revue Historique* who wrote a biography of Michelet, and Camille Jullian, a student and later major interpreter of Fustel. Even Charles Seignobos, who was so closely associated with political history, criticized historical studies in

Germany for being limited to a study of events and for neglecting the study of customs and social and political institutions.[2] A more direct challenge came from the Inspector General of the Libraries and Archives, Paul Lacombe, who openly advocated a more sociological approach to history. An alternative approach was beginning to be developed by economic historians such as Henri Hauser and Henri Sée, though in France economic history still lacked a strong institutional base, being split between the faculties of letters and law. Despite the varying approaches both taken and advocated within the historical profession, it was Durkheim and his followers who would make the most far-reaching attack on political history. Lacking their own *agrégation* and a secure set of chairs, they had the most to gain. They would argue not only that history needed more social content but also that to become truly scientific the field must be remade as a form of sociology, leaving to the historians only the ascientific residue.

The ensuing debate among the French historians and sociologists was not an isolated event. At the turn of the century, many European and American historians began to challenge political history as elitist and to advocate a comparative social and cultural history which would examine the life of ordinary people and analyze anonymous social processes. In place of a linear, developmental approach to history, the new approaches pointed toward historical relativism.[3] These changing approaches also implied shifts in the disciplinary structures. Accordingly, Bernedetto Croce in Italy; Karl Lamprecht, Eduard Meyer, and Friedrich Gottl in Germany; and Franklin Henry Giddings and Charles Cooley in the United States all argued about the appropriate relations between the fields of sociology and history.

The French debate was fueled by these foreign discussions, in part due to the efforts of Berr's *Revue de Synthèse Historique* which published articles by Croce and Lamprecht. Nevertheless, the debate in France was distinctive in that so much was at stake. The historians were firmly entrenched in the Université, holding many positions of power, and the sociologists were still very much on the outside. Adherents of both fields would argue that their field was best suited as an umbrella discipline, encompassing the other. Very much within the bounds of the Nouvelle Sorbonne, the participants often shared *normalien* roots, Dreyfusard sympathies, and even an association with the Bloc des Gauches, a coalition of left and center parties that obtained power in 1903, just as the debate escalated.

For those most to the left, sociology or history transformed into social history seemed the way of the future; some others, however, were more reluctant to change. The participants would debate the respective attention that should be paid to social and political topics, and also how history could become scientific – by adhering to close documentary criticism or by attempting to establish laws through a rigorous use of the comparative method. Not just a discussion of methodological issues and a clash between disciplines, the debate was also colored by generational disputes as young scholars attempted to make their way.[4]

What has become known as the "1903 debate" had in fact significant precursors going back at least to 1888 with the publication of Durkheim's inaugural lecture at Bordeaux. Durkheim had just been appointed as a *chargé de cours* of "social science and pedagogy" – a chair previously identified as simply one of pedagogy and even at that the first pedagogy course to be offered in France (dating from 1882). Other precursors included books by the historians Paul Lacombe, and Charles Seignobos, one of Lavisse's first students at the ENS from the 1874 *promotion* who became a staunch and early Dreyfusard. Now well entrenched at the Sorbonne, Seignobos, in contrast to Lacombe, chose to attack sociology. In 1903 the debate gained full force with an address at the Société d'Histoire Moderne by François Simiand, a follower of Durkheim from the ENS *promotion* of 1893. His sharp and direct criticisms of history were answered by another young scholar, Paul Mantoux (the future author of an acclaimed work on the English Industrial Revolution who entered the ENS just a year before Simiand) and by Seignobos himself.

In subsequent years, the arena shifted to the Société Française de Philosophie, where both Simiand and Seignobos gave papers. In this society, the historians were, for a change, placed in the role of outsiders. The Durkheimians had closer ties to philosophy than any other field, many holding degrees in the subject, whereas most historians had had relatively little exposure to it. In fact, through appointments to the Council of the University of Paris and later the Comité Consultatif, Durkheim came to wield considerable power within the philosophy profession, despite sociology's lack of independence.[5]

The ensuing discussions served to clarify some of the terms of the debate without, however, resolving such fundamental issues as whether the study of events or institutions should take precedence, the meaning of the term social, the relationship between the individual and the social, the appropriate treatment of motives and the subconscious, the role of abstractions, and the degree to which causal explanations and even laws should be formulated. It was a debate which marked both history and sociology for years to come. Although Marc Bloch as a student did not participate directly during these years, he did keep a notebook in which he began to explore some of the issues raised and for the rest of his career he would return to those issues again and again.

Durkheim's inaugural lecture of 1888 questioned both history's method and its scientific status. He asked whether the historians' philological and literary training was a sufficient introduction to the scientific method. Although he acknowledged that history's characteristic role was the study of the particular and not a search for general laws, Durkheim argued that historians nonetheless sought causes and conditions and in the process made inductions and hypotheses. In order to make intelligent selections among the mass of historical facts, the historian needed a guiding idea, and it was this that he could get only from the study of sociology. Failing this, the historian would fall into the trap of useless erudition. In indicating a mutual dependence between sociology and history, Durkheim placed the latter in a subservient role: sociologists were to pose the questions and historians were to provide only the "elements of response."[6]

In 1894, Lacombe argued in *De l'histoire considérée comme science* for a history that was more truly scientific, closely resembling sociology, as defined by Durkheim in his article, "Les Règles de la méthode sociologique."[7] Lacombe advocated a history that looked for causes, made conscious use of induction, deduction, and hypotheses, and sought similarities and generalities through the comparative method. In addition, he argued that history must concern itself with a study of institutions and not just of events and that one should make a distinction between great causes related to institutions and lesser ones related to specific events.[8] In brief, Lacombe appeared to adopt many of Durkheim's positions, but used them to propose a new more scientific history rather than subsume history within sociology. Support for Lacombe's positions soon came from such prominent historians as Gabriel Monod and Henri Pirenne (of the University of Ghent).[9] However, others, most notably the Romanian historian Alexandre-D. Xénopol, challenged Lacombe's positions in the *Revue de Synthèse Historique*. Well versed in the French debates, Xénopol (rector and professor of the University of Jassy) argued that history should be limited to the study of successive phenomena, looking for change and "difference" rather than similarities and generalities.[10]

In 1901, while the arguments between Lacombe and Xénopol were still in progress, Charles Seignobos published *La Méthode historique appliquée aux sciences sociales*, the book which proved to be the catalyst for the 1903 debate. This work was an outgrowth of his course at the Collège Libre des Sciences Sociales, a social science teaching institution which offered an eclectic set of courses for a modest fee, but no degrees.[11] Seignobos belittled the achievements of the social sciences and chose to speak not of what history could learn from sociology, but instead of what sociology had to learn from history. History, he argued, should be viewed as an approach to knowledge, i.e. as an umbrella field rather than a specific science. Since the social sciences depended on historical documentation, they too required the historical method (something which Durkheim did not deny)[12] and could benefit from a "scientific police" such as the *Revue Critique d'Histoire et de Littérature*, which had done so much to enforce the rules of the historical method.

To Seignobos, the efforts to create a science of sociology were premature and likely to fail, given the field's abstract character. Always striving to return to the "concrete reality" of history,[13] Seignobos was convinced that social phenomena were ultimately reducible to individual ones, a position that was directly opposed to one of the most basic tenets of Durkheimian sociology. Furthermore, in opposition to Comte (and also Durkheim, who was not mentioned), he charged that sociological analysis was inherently subjective and that "scientific causes, that is to say the determinant conditions of social facts, are always interior states, motives."[14] On a more conciliatory note, he suggested that the comparative method was "the only procedure with which to establish which phenomena are generally linked together and which are independent."[15] However, he added that such a method rather than being peculiar to sociology required a combination of the methods of the social sciences with those of history.

Despite his concessions to the comparative method, Seignobos made it clear that he saw political history as far more important than social history. The latter, he suggested, was restricted to the study of economic customs, economic doctrines and demography, arguing that these were typically the topics covered by those interested in social science in the past. Social history was seen largely as an auxiliary field which uncovered only necessary conditions and which was heavily dependent on the other forms of history to provide explanations of the phenomena that it claimed to encompass. According to Seignobos, rather than representing the promising new historical orientation as argued by Pirenne, Monod, and Lacombe, "social history is only a fragment of the general history of humanity."[16]

In the meantime, Durkheim and his collaborators had been engaged in what could be seen as sociological policing. Their *Année Sociologique* enforced not so much the rules of documentary criticism which Seignobos held so dear, but instead the tenets of a new, more rational, form of sociological inquiry. In Durkheim's preface to the first volume (1898), he argued that sociology would be built on materials associated with other more established fields – a marked contrast to Seignobos, who had defined sociology and social science by what had been done under its label in the past. History was singled out as a discipline from which sociologists would derive much of the empirical material for their new field. Conceding that sociologists often lacked the expertise to make effective use of the findings of historians, Durkheim argued for an interpenetration of the two fields. In a more critical vein, he cautioned that only those works which could serve as the basis for scientific comparisons were useful, lamenting that historians too often failed to use the comparative method which could take them beyond their particular observations. As a result, sociologists often had to "extract the objective residue" from their works.[17] Among the works which Durkheim reviewed for the new journal were the Seignobos book, reviewed in 1902, and in 1903, an article which also discussed whether or not history could be considered a science, by the Italian historian Gaetano Salvemini (then adjunct at Messina and later well known for his 1905 work on the French Revolution). In these reviews, Durkheim made it clear that he saw the comparative method as specifically sociological and that accordingly scientific history was necessarily a form of sociology and as such could be renamed "dynamic sociology."[18]

More generally, a 1903 article by Durkheim and Paul Fauconnet for the *Revue Philosophique* attempted to establish the character of the relationship between sociology and the social sciences and determined that "sociology could only be the system of the social sciences."[19] Thus, they wanted to place sociology in a position similar to that claimed by Seignobos two years earlier for history. In its role as the leader of the social sciences, sociology was to teach the importance of limiting study within the social sciences to social facts and to help show the separate fields what they held in common. They also stressed the importance of studying institutions and collectivities and of using both the comparative method and the methods of statistics. Even more ambitiously, they argued that "the diverse scientific enterprises discussed here . . . all rest on the principle that social phenomena obey laws and that these laws can be determined."[20]

The most direct and thorough attack on the French historians came from François Simiand, who as a specialist in economics and statistics found himself in conflict with their methods. In the process of completing his doctorate at the Faculty of Law, Simiand was then librarian for the Ministry of Commerce. This position, marginal to academia, allowed him to be very critical of those within.[21] He argued that the historians paid too little attention to economics and also that they approached the study of economic phenomena in a very inappropriate way. Like Durkheim, Simiand directed much of his criticism of history at Charles Seignobos.

Simiand launched his attack at the Société d'Histoire Moderne in January, 1903 – the accepted opening of the "1903 debate." He then published an expanded version of his remarks in the *Revue de Synthèse Historique*, and his comments there set the terms of the debate for many years to come. Berr's journal gave particularly extensive coverage to the early stages of the debate and by so doing no doubt heightened its profile.[22]

The first part of Simiand's paper was a rebuttal of Seignobos' criticisms of social science. According to Simiand, sociology could become a "positive" science, different perhaps in degree but not in nature from other "positive" sciences. "A social fact," he claimed, could be studied objectively, and such a study would require the use of "abstractions," something which contrary to Seignobos need not be feared. Social phenomena were not simply the results of individual phenomena and thus their study could not be broached through an examination of individual motives or agents. Any attempt to explain social phenomena by human motives constituted explanation by "final causes," a form of explanation that should be eliminated from "positive" science. Referring to Seignobos' *Histoire politique de l'Europe contemporaine*, Simiand observed that historians often exaggerated the importance of "contingency" since they directed their attention only to the superficial actions or "occasional causes" and not to "profound causes." Contrary to Seignobos' remarks, the Revolutions of 1830 and 1848 and the War of 1870-1871 were not merely accidental and the result of the actions of a few individuals. Instead, one must study the social disintegration which made these events possible. Once one's attention shifted from individuals to institutions, direct examination of the object of study became possible.[23]

The second part of the paper challenged the discipline of history, particularly as practiced and proselytized by Seignobos and Henri Hauser, who entered the ENS eight years before Simiand and was then professor of history at Dijon. Although more open to social science than Seignobos, Hauser had also asserted that history should not be reduced to "a sort of social dynamic" at the opening meeting of the Société d'Histoire Moderne.[24] History, according to Simiand, could not be considered as a photograph of the past but instead presupposed a conceptual framework, even if it was not totally conscious. The "chronological framework," for example, led historians to assume that there were causal relationships between very diverse phenomena, simply because they preceded or were coincident with particular events. Similarly, the pervasive practice of dividing history into political reigns often resulted in an inappropriate framework for study.[25]

Simiand then proceeded to attack Seignobos' classification scheme for "the essential phenomena of every society." He argued, for example, that the fundamental division made by Seignobos between material and intellectual phenomena was a metaphysical one, since it was the relationships between man and the material world that were of interest. The distinction between public and private was also said to be arbitrary and to make no sense in primitive or feudal societies. Furthermore, in his classification of economic phenomena, Seignobos had included many non-economic topics, such as technology, whereas he had excluded others, such as price and money, which were fundamental. In addition, Seignobos had classified industries according to their end, rather than the materials used, despite the agreement by experts that a materials classification was indispensable. Even Seignobos' adoption of the standard division between agriculture, industry, and commerce was attacked, again because it made sense only for contemporary societies.

Simiand lamented that one could not expect the methods of a group as established as the historians to change quickly. Instead of expecting a quick and total transformation, critics would be well advised to concentrate their attacks on three of the historians' idols: "the political idol" – or the undue emphasis given to political phenomena; "the individual idol" – which implied that history was composed simply of the study of individuals; and "the chronological idol" – which was obsessed with the study of origins rather than a study of "the normal type," and which considered all epochs to be equally important. Of all these idols, the chronological was the most entrenched and would be the most difficult to uproot. According to Simiand, these idols should be replaced by an historical study using "a reflective and truly critical method."[26]

The following year in his own journal, *Notes Critiques: Sciences Sociales*, Simiand continued his attack singling out the program for economic history for the Congrès des Sociétés Savantes to be held that spring. He argued that rather than the indiscriminate detailed studies proposed, more specific studies were needed to answer significant questions that could not be addressed through general documents. According to Simiand, not only was little direction being given for work in economic history but also the historians appeared to have little appreciation of the specific training that was needed for economic history, namely, that of positive economic science.[27]

Simiand's attacks did not go unanswered by the historical profession. The initial reaction from the historians present, including Gustave Bloch (ENS 1868, i.e. six years before Seignobos), was that Simiand's presentation was too abstract and that future discussion should be limited to a "precise point." Simiand agreed to return and speak on the price of coal, but despite prodding from the society, never did.[28] Presumably, he preferred to shape the debate in his own terms and to retain his role as a challenger.

The most immediate published response came not from the established powers such as Seignobos or Monod but instead from Paul Mantoux, who was still working on his doctorate. In an article which also appeared in the *Revue de Synthèse Historique* in 1903, Mantoux argued for greater cooperation between the fields of sociology and history. Since sociology relied on documentation and studied historical phenomena, it required the help of history. The historians for their part should try to orient their research in such a way as to be more useful to the sociologists and so to contribute to the development of the new science. This cooperation would involve abandoning "certain methods of exposition" and "certain traditional classifications, rightly criticized by Simiand." Nevertheless, Mantoux believed that the fields of history and sociology were distinct. The first, despite the assertions of Lacombe, was not a science but instead "a narrative, a description, a tableau," whereas the second was a science seeking causal relations and studying phenomena which were repeated.[29]

More specifically, Mantoux criticized Simiand for his insistence that, *a priori,* social phenomena could be explained only by other social phenomena. In "the concrete reality" of history, institutions and events were, he argued, far too closely linked to allow such a strict separation. On the subject of the boundary between the social and the individual, Mantoux addressed the long-standing debate between Durkheim and Gabriel Tarde (director of the Bureau of Statistics of the Ministry of Justice and in 1900 professor of the Collège de France and member of the Institute). In opposition to Durkheim, Tarde based his form of sociology on the concept of imitation and tried to explain all social phenomena in terms of individuals. Mantoux agreed with Durkheim that one should draw a distinction between voluntary and obligatory imitation (or social constraints), but he also suggested that the first might lead to the second and that the possibility of such a process ought to be examined historically. According to Mantoux, "It is possible – even probable – that social phenomena have only other social phenomena as a cause, in the scientific meaning of the word. But that is only a presumption that one must admit provisionally and conditionally."[30]

In a related argument, Mantoux dismissed Simiand's fear of "finalist" or teleological explanation as unfounded. According to Mantoux, there was no reason why an explanation by motives should be incompatible with "positive" social science. Legitimate concerns over teleological explanations in the physical sciences stemmed from an inappropriate anthropomorphism that transferred explanations from one domain to another, but in social science such an argument was not valid. If sociology were to allow a study of motives, its need for history would be clear. History could provide information not only about the conscious aspirations of society but also about unspoken needs, vague opinions, and changing customs and values.[31] In answer to Mantoux's charge, Simiand argued the following year that "finalism" (i.e. teleology) could not be reduced to anthropomorphism. Furthermore, anthropomorphism could be very dangerous for social sciences particularly when formulated not simply as "things are made for man" but more subtly as "things are made as I, a man of a particular culture and particular status, feel that they must be."[32] Simiand remained convinced that sociology must follow the strict rules of "positive" science.

In a more general argument, Mantoux suggested that in striving for scientific status, sociology should not make the common error of borrowing its methods uncritically from other very different sciences. As a model to follow, Mantoux suggested that rather than turning to the physical and natural sciences, sociology should look to linguistics. That field had passed through a useful period of "historical preparation" (without becoming confused with history), had developed its own methods and laws appropriate to its object of study, and had looked at both individual and social causes.

Seignobos' most immediate answer to Simiand's attack came in December 1903, during a lecture series at the Ecole des Hautes Etudes Sociales on the "Relationships of sociology with the various social sciences and auxiliary disciplines" that was organized by Durkheim. Modelled on the Ecole Pratique des Hautes Etudes, this institution offered seminars by leading members of the Université and provided a forum for issues not treated there.[33] Given the setting and his topic (the relationships between sociology and history), Seignobos could not totally ignore the attacks to which his position had been subjected. Nevertheless, he maintained his earlier rather condescending attitude toward the sociologists and chose to ignore many of their objections. Predictably, the thrust of his talk was to insist that sociology required the help of history as it needed both to study the phenomena of the past and to use the historical methods of documentary criticism, which he and Langlois had promoted in their historiography course. Addressing the other direction of the relationship, Seignobos admitted that the historian needed to classify the historical facts. Since historians increasingly chose to study economic phenomena, sociology could aid in establishing a methodological framework, becoming, he noted somewhat patronizingly, an auxiliary of history. In a more significant concession, Seignobos noted that in their search for the causes behind social evolution, the historians not only needed to use the comparative method but should turn for help in this matter to sociology.

Examining the boundaries between the two fields, Seignobos observed that sociology did study some of the same phenomena as certain subfields of history, such as the histories of language, religion, and social organization, but then only when those fields were concerned with a study "of masses" rather than with geographical and chronological detail. Here Seignobos implied a somewhat broader definition of sociology than that included in his 1901 book, which had excluded religion, language, and much of social organization (such as the family, property relationships, and social class) from both social history and sociology. On the other hand, he remarked that as sociology did not cover phenomena resulting from conscious human acts and individual acts, it excluded the arts, literature, politics, and part of economics, all of which were covered by history. Such phenomena required an examination of motives and the use of intuition and analogy, admittedly a somewhat inferior method but the only one appropriate to their study. In answer to suggestions for a "unification" of the vocabularies of history and sociology, Seignobos argued that historians would be better advised to maintain their established vocabulary, which was comprehensible to all, and let the sociologists develop their own particular terminology. The sociologists, after all, could not even agree amongst themselves. Clearly, in Seignobos's eyes, Durkheimian sociology was in no position to subsume history since it covered only a fragment of its domain and since even within the fledgling field of sociology, the Durkheimians had yet to achieve dominance.[34]

Following a hiatus of two and a half years, the series of debates at the Société Française de Philosophie began. They focussed first on the issue of causal explanation and later on the subconscious. Simiand began the discussions with a paper on causality in history. He charged that Seignobos, in his *Histoire politique de l'Europe contemporaine* (1897), was often content simply to record rather than to explain and that he fell far short of establishing "a valid law of the succession of phenomena." To direct historians onto the right path, Simiand proposed four rules: (1) "define the precise effect in general terms"; (2) "select as a cause the antecedent which is linked to the effect by the most general relationship"; (3) "specify the immediate antecedent"; and (4) "in the research for causes, search for propositions for which the reciprocal is true." In the discussion that followed, Lacombe remarked that he could identify with Simiand's desire to exclude the individual from study since the individual was associated with the contingent, but he also argued that such a task was impossible. Simiand replied that his aim was not to exclude the individual, but simply first to try to explain as much as possible by "regular causation." One could, he claimed, explain far more in social terms than was generally thought.[35]

Seignobos, who was not present at the 1906 meeting, gave his reply the following May in a paper on the practical conditions affecting the search for causes in historical research. According to Seignobos, history, which was a "fragmentary and indirect study of phenomena of the past"[36] was incapable of establishing laws, and furthermore, even explanations were often impossible. More specifically, in response to the rules of work formulated by Simiand, Seignobos noted that defining the precise effect in general terms went against the particular character of historical explanation, and that it was also impossible to achieve, given the complexity of historical events and the state of historical knowledge. Similarly, the historian was in no position to identify the antecedents tied to an effect by the most general relationship; instead historians would be better advised to label as cause "the most immediate antecedent" linked to the phenomena of interest by "the most particular relationship," since that was the only one he knew for certain. Although the rule demanding that one specify the "immediate antecedent" had value, unfortunately even this was not always possible. Finally, the historian knew far too little of causes and effects to formulate propositions for which the reciprocal was also true.

In short, Seignobos felt that Simiand had a very unrealistic view of historical enquiry, a result in part of Simiand's philosophical training. Whereas philosophy operated at the highest level of science and sought to define the character of an ideal science, history represented the most imperfect form of knowledge at the bottom end of the scale, given the fragmentary character of its material, its use of indirect observation, and its need to reason by analogy. Nevertheless, despite its imperfections, history, according to Seignobos, had a special value to counteract the negative effects of overspecialization:

It seems to me that the main usefulness of history is to remind us of the conditions of existence of reality, to show us that there is never a parallel series of phenomena, but that everything is continually intertwined. History serves to make us understand that each phenomenon is the result of many phenomena of extremely different types which occur in different places; it reminds us of the complexity, the interdependence, which connects different "pays" and series of different phenomena.[37]

As a view of an umbrella science, this conception differed significantly from that proposed by the sociologists for their field. Rather than providing directing hypotheses and establishing logical relationships between carefully defined and structured subfields, history was simply to show an empirical interconnectedness between a great variety of phenomena.

Several of the other historians who were present criticized Seignobos for his restricted view of history. Opening the discussion, Gustave Bloch argued that for certain very simple historical phenomena, such as man's need to settle near water sources, one could establish laws though even then there would be exceptions. In turn, Gustave Glotz (from Hauser's ENS *promotion* of 1885 and then a *chargé de cours* at Paris) pushed Seignobos to admit that one should ask questions, search for explanations, and even "laws of general scope." Seignobos' answer would later come to symbolize the limitations of his form of history; he would only concede, "It is very useful to ask questions, but very dangerous to answer them." Among the other observations offered by the historians, Lacombe criticized what he saw as Seignobos' very narrow conception of history as the study of events rather than of customs or of institutions in the large sense. To this, Seignobos answered that customs could only be understood after a careful scrutiny of events.[38]

The bulk of Simiand's remarks were devoted to a discussion of the terms "cause" and "condition." Countering comments by Seignobos, he argued that the distinction between the two was indeed of considerable importance. He also emphasized the need for comparison, for the use of abstraction, and for a technical language to replace that of everyday, which was riddled with ambiguity. Finally, Simiand took exception to Seignobos' statement, "Sociology, in the sense of Comte, tends to remove every psychological factor for the explanation of phenomena – and it seems to me that Simiand is of the same opinion." Although admitting that he was against an explanation stated in terms of goals, ends, and motives, Simiand explained that it was only subjective psychology which he sought to eliminate, not objective, and that in fact he saw objective psychology as "the essence and the character of the phenomena which we study and the causes which we seek to find." Elie Halévy stressed this distinction between subjective and objective psychology and, noting that it was one Durkheimian principle that many failed to understand, suggested that a later meeting be devoted to its discussion.[39]

That meeting, the last in this series, took place the following year (1908) when Seignobos presented his paper on "The Unknown and the Subconscious in History." The title alone indicated that he had failed to grasp Simiand's distinction. There Seignobos insisted that "phenomena became more intelligible and knowable as the ideas and conscious causes play a larger role" and that social phenomena could only be understood in terms of individual ones.[40] In the wide-ranging debate that followed, the sharpest exchanges took place between Durkheim and Seignobos.

Durkheim insisted that one must study the subconscious and that in fact it posed no special problems. With the appropriate comparative methodology, both the subconscious and the conscious could and should be studied by the two interconnected fields of history and sociology. Stressing the necessary similarity between the fields, he claimed, "In truth, there is not to my knowledge, a sociology which merits this name which does not have an historical character."[41] Furthermore, Durkheim argued that not only should the historian look for causes but that "Each relationship of causality is a law." Durkheim also took Seignobos to task for arguing that events were easier to understand than institutions; only by understanding the "framework of a society" could one come to understand events.[42]

For his part, Seignobos continued to insist that the conscious testimony offered by witnesses formed the most useful evidence for the historian, that the comparative method could do little, that at most the comparisons used by historians were simply analogies, and that events were far easier for the historian to understand than institutions. On the issue of causes and laws, Seignobos argued that there is "no certitude in history." He had, in short, reverted to his early position retracting even those few concessions on causal explanations and the importance of the comparative method which he had made when speaking in Durkheim's forum at the Ecole des Hautes Etudes Sociales.[43]

Gustave Bloch sided with neither Seignobos nor Durkheim. Of Seignobos, he remarked, "I am really frightened by the skepticism of Mr. Seignobos. According to him, what remains of history? Practically nothing." Attacking Durkheim's arguments that history involved a search for laws and that the method of the biologist and the historian were really quite similar, Bloch clarified how his position on the use of laws in history differed from that of the Durkheimians. As repetition was impossible in history, one could not, as Durkheim had suggested, equate causes with laws. Laws in history were only possible for very simple and crude phenomena such as those of *géographie humaine* but not for psychological phenomena which were diverse and complex.[44]

Such a view did not coincide with that of Durkheim. He retorted, "Then one must give up formulating causal relationships." Célestin Bouglé, an active member of the Durkheimian group,[45] concurred, arguing, "I believe, like M. Durkheim, each causal explanation, to really be an explanation, must refer to laws." Historians, he continued, formulated such laws despite themselves, citing an example from Gustave Bloch's own work in which he had stated that the system of protection "dominated whenever the state was incapable of assuring the security of individuals."[46] Bloch's defence was simply that he and others needed to be more careful and to phrase their statements with more qualifications. He was not willing to adopt as rigid a view of the science of history as that proposed by the sociologists.

During the course of this debate, the participants frequently approached the issue of "collective consciousness," but Durkheim insisted that that was a separate issue, which he did not wish to discuss.[47] Seignobos appeared to be very skeptical of the value of this concept and in fact anything that appeared to be "social." In closing, André Lalande, a philosopher who chaired the meeting, tried to find some common ground between them, allowing at least a limited role for social conceptions – but without much success. He suggested that both agreed that individuals can never exist before and outside of society:

> Mr. Durkheim: Let us rest with this illusion and say that Mr. Seignobos admits, as I do, that *le pays* changes the individuals.
> Mr. Seignobos: So be it but on the condition that *le pays* is only conceived as the ensemble of individuals.
> Mr. Durkheim: Let us say, if you prefer, that the assembly changes each of the elements assembled.
> Mr. Seignobos: I will admit this tautology.[48]

By 1908, after the debate between the historians and sociologists had raged for several years, Seignobos had become increasingly entrenched; Gustave Bloch started to clarify just how he differed from the sociologists; and Durkheim became harsher in his criticisms of Seignobos. Although the historians continued to argue among themselves over the fundamental character of their discipline with Seignobos taking the most extreme position, the sociologists maintained a more unified front, making no significant concessions. They continued to insist that social phenomena could be explained only by social facts, that teleology and explanation by motives and goals must be avoided, that psychological phenomena examined by sociologists could be studied objectively, and that sociological laws true for all cases covered by the law could be formulated. For the young historians, the terms of the debate became clearer, but the answers perhaps less so.

At the time of the debate, Marc Bloch was still a student and as such not an active participant in the published discussions. However, as a student of both Seignobos and Durkheim, not to mention his father Gustave, he was directly confronted with the arguments, and as a *normalien*, he undoubtedly would have followed the debate in its published form. Parts of his 1906 notebook entitled "Historical Methodology" do support established history as represented by Seignobos. He noted, for example, that the historical method was descriptive and chronological rather than analytical and also did not have a scientific existence. Accordingly, the form of inquiry used by the historian and the chemist were very different. On the other hand, Bloch was critical of such an historical method and apparently agreed with several of Simiand's arguments. Bloch argued that history should not simply collect experiences, but needed to interpret them. Also, when studying what he termed "psycho-social phenomena," one should, according to Bloch, replace the chronological and empirical methods with an analytical one.[49] Similarly, he remarked that the notion of "event" ought to be replaced with that of "phenomenon" – which was "the product of the analysis of the event."

On the issue of the separation of the individual and the social, Bloch appeared to side with Mantoux and Lacombe in thinking that it was impossible to separate the two, and claiming that such an attempt was misdirected. Although using the same phrase that Seignobos would later use in answer to

Durkheim, "society is only an ensemble of individuals," Bloch quickly modified this position in a discussion of religious phenomena. A social phenomenon in religion was said to be a grouping rather than an ensemble of individual religious phenomena, a group which could have its own laws. However, in contrast to the Durkheimians, Bloch argued that such laws were superimposed on individual laws, indicating not only a possibility for individual laws (something with which both Durkheim and Seignobos would disagree) but also an interplay between the individual and social that went against Durkheimian principles.

Even though Bloch saw a possibility for laws, both social and individual, he claimed that the only "human science" which currently was in a position to catch a glimpse of such laws was philology. Appearing to contradict this statement, he later foresaw the possibility of "economic laws." He restated his father's example in economic terms by noting, "In all rich *pays* where the water sources are scarce, population is grouped in large villages," and also formed a corollary relating settlement in large villages to the form of property.[50]

According to Bloch, not only could one not separate the individual and the social, but also one could not separate psychological and social phenomena. This indicated that he at least understood the psychological character of social phenomena as conceived by the Durkheimians, but did not directly address the issue of a distinction between objective and subjective psychology or the related questions of teleology and motives. He did however appear quite intrigued with the subjective elements of perception that the Durkheimians would consider outside their realm.

Another distinction which Bloch failed to address in his early notes was that between sociology and history. At one point in his notes, he even changed the word "historian" to "sociologist." There he wrote, "A sociologist (historian) writing a book on Louis VII and his times makes me think of a physician who would record the results of his research in an octavo notebook entitled 'My assistant, Mr. so and so, and my laboratory.'"[51] He also argued that one should concentrate on the subject or core of a study rather than choosing such "inevitably absurdly complex and poorly deliminated" topics as Louis VII. It seems that not only did he see history and sociology as very similar but that he preferred the sociological method to that of many historians. Not unlike Lacombe, he suggested that one should search for generalizations and relate one's findings to proposed laws, and, like Simiand, he advocated greater coordination of research even to the point of using questionnaires. To begin a study of the "history of societies," Bloch suggested that one first examine that which was most determined, namely language and economic phenomena – two areas which would color much of his own research and which were of much interest to the Durkheimians.[52]

In this tentative statement while still at the ENS, Bloch demonstrated not only a need to confront the issues of the "1903 debate" but also considerable sympathy with the positions of the Durkheimians. He did not, however, show a willingness to dismiss the individual from study. Furthermore, he avoided some of the more contentious issues relating to objectivity, teleology, and motives and the difference between sociology and history. In addition, he remained unclear both on the extent to which one could formulate laws in the *sciences humaines* and on the meaning of the term "social phenomenon." These notes proved to be just a first stab at a nagging debate, as Marc Bloch would return to the questions raised again and again.

4

The quest for identity in Vidalian geography

The differences between this geographical science and the purely human sciences, such as sociology and history, result clearly enough from the explanations that we have given that it would be superfluous to emphasize them. Though of different orders, they [these sciences] are called upon to render great services to each other; however it is essential that each remains keenly aware of its object and its own method.[1]   Vidal de la Blache, Revue de Synthèse Historique, 1903

In their quest for legitimacy, French sociologists challenged not only the established field of history but also the relatively new field of human geography as represented by Vidal de la Blache and his followers. The claims both for scientific status and for a distinctive object of study for human geography were questioned – particularly when the geographers seemed to impinge on material that the sociologists could claim for their own. Once again, the discussions took place within the confines of the Nouvelle Sorbonne among academics who shared *normalien* backgrounds and staunch support for the Republic.

Though his active political involvement was somewhat limited, Vidal, like many of his historical colleagues devoted much of his pedagogical and academic work to the creation of a renewed and strengthened vision of France.[2] Vidal had attended the ENS nine years before Seignobos and thirteen before Durkheim and in fact was already teaching at the ENS when Durkheim arrived as a student. He was therefore senior to many of the participants in these discussions, but still faced a struggle to gain acceptance for geography as a separate professional field.

In contrast to the "1903 debate," the ensuing discussions between geography and sociology failed to take the form of a direct debate. There was no single treatise on the order of the books by Lacombe and Seignobos to lay out the geographers' positions and no common forum such as the Société Française de Philosophie where the key actors were forced to confront each other. Instead, the attack by the sociologists was less direct and more piecemeal, and the Vidalian position, to the extent that it can be identified, was often both ambiguous and vacillating.

Initially, the confrontation between geography and sociology came indirectly as both Vidal and Durkheim attempted to define their respective domains. With time the sociologists attacked geography more directly and even argued that the field of human geography should be transformed into social morphology and subsumed within sociology. However, rather than addressing Vidal, they often chose to criticize the German geographer Friederick Ratzel whose field of *anthropogeographie* had helped to inspire Vidal's *géographie humaine* but was certainly not identical to it.[3] The discussion was also somewhat confused by the attraction of some of Vidal's students to the Durkheimian ideas, and in a couple of cases their contributions to the Durkheimian journals.[4]

Seemingly serene in their newly acquired status, Vidal and his followers would not explicitly address the sociologists' criticisms. Occasionally they would make passing and oblique references to the criticisms in their methodological articles, addresses, and reviews, but more often they just carried on with their substantive work, which was often promoted and praised by members of the historical profession. Though still institutionally tied to history, the Vidalians succeeded through their substantive work in both improving their field's academic reputation and acquiring key positions in the Université. The most direct confrontation between sociology and geography did not come until 1910, following praise and appointments for the Vidalians. Then, Simiand attacked a series of regional monographs, written as doctoral dissertations under Vidal's supervision, that had become the cornerstone to the new field. Vidal's response was characteristically muted and even more unclear than before.

Both trained and intrigued by the early Vidalian geographers, Marc Bloch was, nevertheless, very open to the criticisms directed at them by the Durkheimian sociologists. Reflecting the indirect and somewhat delayed confrontations between sociology and geography, he did not explore the questions raised by them in his 1906 notebook on historical methodology. Their centrality to Bloch's particular and evolving view of history, however, become apparent in his later substantive work and his reviews.

Vidal's early depiction of a scientific geography can be found in the *Annales de Géographie*, begun in 1891. In their opening issue Vidal and Dubois suggested that the time was now ripe for a scientific journal of geography in France. Rather than simply enumerating discoveries, the journal was to link new findings to earlier ones and develop a method.[5] It was not until four years later, however, that Vidal attempted to elaborate on that method. He suggested that the geographer's task should not be limited to the description of local areas but instead should also include classification and even explanation. Not content just to disclose particular causes, he promoted a form of explanation based on general elements leading ultimately to the establishment of geographical laws. Underlying his rationale for a geographical science was the concept of "terrestrial unity." To Vidal, this implied that similar processes acted throughout, that comparisons could lead to general explanations, and that phenomena in one part of the world were influenced by those elsewhere.[6]

Another central concept for Vidal was that of *enchaînement* or linking. In an 1888 article in the *Bulletin Littéraire*, he argued that the goal of geography was to explain phenomena by *l'enchaînement* between them, and for him, the crucial linking was that which occurred between different orders of phenomena. Ten years later in the *Annales de Géographie*, he praised Ratzel's attempt in *Politische Geographie* to link the phenomena of political geography and physical geography through the geographical study of plants. Such a linking, Vidal suggested, could do much to explain population distributions and the like. He also endorsed Ratzel's depiction of the world as a "living organism" since that concept could highlight the processes of change.[7]

By the time of Vidal's opening lecture at the Sorbonne in 1899, he felt in a position to proclaim the success of the effort to renew geographic studies. Vidal, himself, had been a key actor in this effort both at the Faculty of Letters at Nancy and even more actively at the ENS, to which he was appointed in 1877. His statement, in part a response to the growing challenges from sociology, demonstrated that his conception of this new science was not altogether clear. As in his earlier article on Ratzel, he noted that with geography's advance, other sciences had become more geographical including not only the "sciences of the earth" but also "certain sciences of man."

To highlight geography's distinctiveness, Vidal argued that its particular task was that of explaining the diverse milieux of the earth, and accordingly, he suggested that the time was now ripe for an emphasis on regional studies. Such research could lead to explanations of the *contrées* (lands or regions) that were both "descriptive and reasoned" and disclose not only the action of general laws but also the modifications caused by local circumstances. Despite the emphasis on "concrete realities," Vidal argued that the regional investigations would allow one to determine the more general interconnections between "the soil, the sky, the plants, and the human works."[8] Unwilling to abandon his goal of scientific status, Vidal remained caught between the conceptions of geography as a general science and geography as regional study.

Without a doubt, sociology was one of the "sciences of man" which prompted Vidal in 1899 to attempt to specify geography's distinctiveness. Durkheim and his associates had launched their *Année Sociologique* in 1898 and included in the first issue a section entitled *la socio-géographie*. There, a review of Ratzel's *Der Staat und sein Boden geographisch beobachtet*, generally attributed to Durkheim, both criticized his method and suggested that sociologists could learn from the geographer to pay more attention to the "territorial factor." That factor, it was argued, could affect certain "collective representations," the general structure of societies, and the movement of people. Attacking Ratzel's treatment, the review claimed that his theory, which was poorly expressed, rested on some very debatable principles such as the composition of societies by autonomous individuals, and furthermore that in looking at the relationships between societies and the soil, one must not forget that it was the social content of a changed physical geography which was most important.[9]

The following year Durkheim suggested that one of the tasks of the sociologists was to provide a new organization and classification of knowledge as related to social phenomena. To illustrate the need for new disciplinary divisions, he pointed to geography which was only beginning to establish links with demography, a science with which it had so much to share.[10] Accordingly, the sixth section of the *Année Sociologique,* which had formerly included *l'anthropo-sociologie, la socio-géographie,* and questions of demography, was to be reorganized into a section of social morphology. Not only were the sociologists to learn from geography, they were to incorporate it within a new, more rational, division of knowledge.

In his introduction to that new section Durkheim gave a statement on the character of social morphology, which was also an attack on geography as a discipline. The object of study of social morphology was to be the substratum on which social life rested. This substratum included the number of people and their distribution and the nature and configuration of "things of all kinds which affect collective relations" – a category which appeared to include the built environment, political boundaries, and some elements of physical geography. As such, the section would draw on research formerly done in the fields of demography, history, and geography, explaining, for example, the variations in political areas and the development of urban centers. Contrasting these intentions to the existing field of political geography, Durkheim argued, "It is a question, in fact, of studying not the forms of the soil, but the forms which societies take on as they establish themselves on the soil, which is very different." The word "geography," he claimed, would almost inevitably lead one to exaggerate the importance of physical features, such as mountains and waterways, for the social substratum at the expense of such crucial factors as the number of people, the manner in which they are grouped, and the form of their dwellings.[11]

In later reviews of Ratzel's work for the new section, Durkheim criticized geography directly for attempting to explain societies by physical geography. His 1899 review of *Politische Geographie,* the same work praised by Vidal, charged that despite some similarities between Ratzel's depiction of political geography and social morphology, Ratzel still attempted to determine the influence of physical geography on the political development of peoples. Since, according to Durkheim, social phenomena could only be explained by other social phenomena, this explanation by physical features was unacceptable. Instead, societal forms needed social explanations, such as those of the state of the economy and the spread of religious ideas. Durkheim fell short, however, of totally ruling out a role for the elements of physical geography in such a study:

It is possible that the findings of physical geography also have a role in this matter, but they are only one of the causes that help to produce the phenomena studied. They become, by contrast, the essential and practically only cause considered if one's object is defined exclusively as the way in which they affect the development of the state.[12]

The key phrase here appears to be "essential cause" – something which was inadmissible given Durkheim's view of social explanation. He was, however, willing to admit that features of physical geography might enter into the picture as one of the contributing factors, meaning perhaps something on the order of what Simiand labelled "conditions" in his debate with Seignobos over the meaning of the term "cause." Vidal, by contrast, appeared to have no problem with more direct links between physical features and social phenomena.

In a review in 1900 of the first volume of Ratzel's *Anthropogeographie*, Durkheim clarified that even in their contributary role, physical features could only come into play for relatively minor phenomena. Whereas the "terrestrial influences" could contribute to the formation of "the idiosyncrasy of peoples," they could not account for even one of "the constituent features of social types."[13] Despite Durkheim's concern that sociology become more empirical so that it could "go beyond generalities,"[14] he had clearly not meant a mixing with the "concrete reality" to the same extent as that advocated by Ratzel.

It is interesting that Ratzel was singled out by Durkheim as he attempted to promote social morphology. The French geographers, by contrast, received only cursory treatment. In a short unsigned note, Durkheim praised Vidal's largely laudatory discussion of Ratzel; Vidal's main criticism had been simply that some of Ratzel's propositions were too dogmatic.[15] According to Durkheim, Vidal presented a more precise depiction of political geography than that to be found in Ratzel's work.[16] By contrast, Durkheim reserved harsher treatment for Franz Schrader, the cartographer for the publishing house Librairie Hachette and for Edmond Demolins, who did not hold a university post and was a follower of Le Play rather than of Vidal.[17] Both were criticized for exaggerating the role of physical factors for societies. Perhaps Durkheim, who still sought to obtain a foothold for sociology within the university establishment, felt that Vidal, long time teacher at the ENS and now holding the chair of geography at the Sorbonne, was too powerful a target, or perhaps he feared alienating some of the students who were so taken with Vidal's teaching. In any case, his criticisms could have applied also to some of the work of the Vidalians.

Durkheim was not alone among the sociologists in criticizing geography. Marcel Mauss and Paul Fauconnet, both relatively young scholars in their late twenties, joined the attack in their 1901 article for *La Grande Encyclopedie*. Echoing Durkheim, they claimed that some of the questions treated by geographers, such as boundaries, routes of communication, and social density, were more appropriate for the field of sociology, as they "stemmed from the nature of societies."[18] Simiand included a "methodological note" implicitly criticizing geography in *Notes Critiques: Sciences Sociales* in 1903, the same year as his famous attack on history. There he asked, rhetorically, whether social morphology was not more logically constituted than *géographie humaine*, using, interestingly, the Vidalian term which had gained more acceptability.[19] In his view, geography was simply a source of general documentation and not a scientific field.

Social morphology was not a major focus of Durkheimian work in the period leading to the First World War. In 1906, Marcel Mauss (with the colloboration of Henri Beuchat) published the only pre-war monograph on social morphology: "Essai sur les variations saissonnières des sociétés Eskimos: Etude de morphologie sociale."[20] They argued that geographers had a tendency to concentrate on the factor of land rather than other aspects of the "material substratum" and often ignored the mediating effects of other factors. For example, a concentrated population resulted not only from characteristics of both the climate and land but also from their moral, legal, and religious organization. The social context and the repercussions of physical factors on collective life were particularly important, both of which were more sociological than geographical. In their monograph, Mauss and Beuchat examined the dependence of social life on seasonal variations in social morphology and suggested that their case study pointed to a law of social life that was "probably of a very broad generality." As they elaborated, "Social life does not remain at the same level at different times of the year; it goes through regular and successive phases with an increasing and decreasing intensity, phases of rest and activity, of expenditure and of recovery."[21]

Discussions of the relations of social morphology with the rest of sociology were relatively few. In their 1901 article, Mauss and Fauconnet had downplayed the significance of the social substratum arguing, "it [sociology] assigns a preponderant role to the psychological element of social life, beliefs and collective sentiments."[22] Sociology, they claimed, was not a materialist discipline. They also argued, however, that one should not attempt to make a strict separation between the phenomena of social structure (as studied by social morphology) and collective representations: "These are inseparable phenomena between which there is no reason to establish a logical or chronological primacy."[23] In a 1909 article on sociology and the social sciences, Durkheim addressed the same issue noting that social morphology studied the external aspect of society and looked at its "anatomy" as opposed to its "physiology" (a topic covered by religious sociology, *la sociologie morale, la sociologie juridique,* and economic sociology).[24]

In short, social morphology differed from *la géographie humaine* not only in devoting more attention to population density and movement but also in intent. Unlike geography, it was not meant to constitute an independent field but instead was viewed as an integral part of the broader field of sociology. Rather than examining a simple relationship between society and environment, it was to be an examination of the structure or "anatomy" of the society, intimately related to its "physiology." Within that structure, geographical factors had a part to play, but only insofar as they influenced the forms that societies took on as they established themselves on the land.

The responses by the Vidalian geographers to these various challenges were rather muted. As in the "1903 debate," Simiand was answered most directly by a doctoral student, rather than an established professional. This time the most explicit answer came in Antoine Vacher's 1903 review of de Martonne's *La Valachie* for the *Notes Critiques: Sciences Sociales,* the same review which Simiand introduced with his methodological note, challenging the basis of *la géographie humaine.* Caught in the middle, Vacher was preparing a thesis in geography under Vidal and was also associated with the Durkheimians having written since 1901 for Simiand's journal, for which he reviewed works of history, geography, and economics. Vacher's *promotion* at the ENS was 1895, only two years after that of Simiand.

Describing human geography as the synthesis of knowledge from natural sciences and historical and economic knowledge, Vacher argued that it had the "right to set up itself as an autonomous discipline and to claim a role in the study of the life of human groups." From de Martonne's book, Vacher claimed, sociologists could note that:

the soil, the climate, the water regime, and the vegetation which results are phenomena from which man does not know how to escape; consciously or not, each human group is subject to their influence which it can, in truth, by its own reaction limit. This cause does not act alone or in isolation. It combines with historical causes of all kinds. But a sociological explanation which strives to be complete can not neglect it.[25]

Human geography, he argued, was devoted to untangling and demarcating "this influence."

In this statement, Vacher was not clear whether human geography was to deal with "man" or with "human groups" – potentially a very important distinction. He also was unclear as to what was the object of the "influence" from a geographer's perspective, i.e. was the influence being studied by human geographers only that of material features on society or were other objects also to be included? It appears that for Vacher, it was the relationship rather than the end-product or object that was of interest, whereas the Durkheimians were clearly interested in an object – social phenomena. He also seemed to foresee no problem in "synthesizing" knowledge from the natural sciences, history, and economics, implying that the findings of these very different fields could be taken as is. For their part, the Durkheimians were far more skeptical as to the ease with which such a synthesis could be made.

Vidal's responses to the Durkheimian criticisms were even less direct and less clear than Vacher's. They can be found scattered in four articles published between 1902 and 1905. In the first, published in the *Annales de Géographie*, Vidal took a rather non-committal position that differed little from his earlier statements on the nature of geography. Noting that the relationship between geographical and social phenomena must be observed from both directions, he stressed the importance of what he termed "general causes" including position, area, climate, and social constraints. Furthermore, although keeping his interest in *la géographie générale,* as shown for example in the establishment of types of civilization, he noted the "peril" of premature generalization and suggested that the best remedy would be "the composition of analytical studies, of monographs where the relationships between geographical conditions and social phenonema would be envisaged up close on a well chosen and restricted field."[26]

In 1903, following Simiand's attacks on Seignobos and his criticisms of geography, Vidal published an article in the *Revue de Synthèse Historique* which attempted both to separate *la géographie humaine* as practiced in France from the anthropogeography of Germany (which Durkheim had attacked), and also to show how *la géographie humaine* differed from "the purely human sciences" such as sociology and history. To illustrate the distinctive character of his field, Vidal argued that its methods, aims, perspective, and origin were very much like those of "botanical geography" and "zoological geography" starting with "the general facts of distribution" and then addressing the questions posed by those distributions. Man, he suggested, should be viewed as part of "the living creation," as a collaborator with nature rather than an adversary. The relationship between man and his environment was said to be very much like that of certain "social plants which make existence so difficult to all others that they take no time in taking possession of large surfaces, only tolerating a few parasitic species."[27]

Despite his rhetoric, Vidal's use of biological and ecological science to provide both scientific status and a separate identity to geography remained *ad hoc* serving to legitimize the field rather than as a specific paradigm or program of research, as Robic has so convincingly argued.[28] Even as a legitimizing device, his use of biology was not altogether successful. For example, the distinction that Vidal attempted to make between *la géographie humaine* and anthropogeography remained somewhat obscure since he noted that men such as Carl Ritter and Ratzel also drew on "the biological method." The main difference seemed to lie more in what he termed "a clearer view of the method and in the perfection of the instruments of study" stemming in part from advances in cartography, statistics, and ethnography.[29]

With respect to the distinction between sociology and *la géographie humaine*, Vidal claimed:

Human geography thus merits this name because it studies the terrestrial physiognomy modified by man: that is why it is *geography*. It only contemplates human phenomena in their relationship with this surface on which the multi-faceted drama of the competition of living beings takes place. There are thus social and political phenomena which do not enter into its competence or which are only indirectly connected to it and which it does not treat. Despite this restriction, it keeps numerous points of contact with this order of phenomena. However, this branch of geography proceeds from the same origin as botanical and zoological geography. It is from them that it [human geography] takes its perspective. The method is analogous; only much more delicate to manipulate, as in every science where human intelligence and will are at work.[30]

Even though Vidal claimed that his object was the terrestrial physiognomy, Vidal still attempted to explain human phenomena by their relationship with the terrestrial physiognomy, arguing, for example, that climatic change in the quaternary epoch influenced the first formations of societies. Also, despite his claim that geography was only to look at human phenomena from the perspective of their relationships with the earth's surface, he would bring in all sorts of other social factors, such as the role of imitation, innovation, and commerce in his discussion of complex societies. Vidal thus left himself open to the basic criticisms made by Durkheim, Mauss, Fauconnet, and Simiand – criticisms which he did not answer. Instead, he was content to conclude that the distinctiveness of his field was self-evident, given the arguments he had made and that it would be "superfluous" to emphasize the differences between "geographical science and the purely human sciences such as sociology and history."[31]

The following year Vidal participated in Durkheim's series of lectures "Sociology and the Social Sciences" at the Ecole des Hautes Etudes Sociales. Speaking on the "Relationships of Sociology with Geography," Vidal attempted to clarify the distinction between these two fields. His most specific explanation came in a discussion of the case when an isolated society came in contact with other peoples. In such cases, he admitted that the social and economic consequences, such as a change in the demand for labor, lay outside of geography's domain. Again, he suggested that the geographer was, by contrast, interested in the ensuing changes to the terrestrial physiognomy. As he elaborated, "these phenomena are geographical as long as they modify the composition of life, the appearance of the surfaces, and the number and relationships of human groups."[32] In so arguing, Vidal included the demographic information which Durkheim held so central for a study of social morphology, without providing a direct answer to Durkheim's charge that such information was in no way geographical.

Vidal's somewhat confusing conclusion was that, "Human geography recognizes itself as a part of the study of the earth and must, for that reason, remain distinct from the sociological sciences. It proceeds from the earth to man, not in the opposite direction."[33] In spite of his attempt to distinguish geography from sociology by its object of study, what Vidal was saying here was that the difference actually lay in the point of departure, respectively the earth and man. Even more significantly, his argument implies that it was a legitimate scientific undertaking to study human societies from the perspective of physical and biological phenomena – clearly something which Durkheim and his associates opposed. In all, Vidal's "answer" remained rather unclear and confusing, and, despite the forum, he failed to make direct references to the specific charges which the sociologists had made of his field.

Similarly, in a 1905 address at the Musée Pédagogique, Vidal argued that geography should "consider human societies in relationship to the earth" – relationships that were reciprocal. In opposition to the Durkheimians, he suggested that geography could be of service to both sociology and history precisely because it examined human phenomena in relationship to natural causes and placed them in their physical and biological milieu. To him this linking of phenomena of different orders was not problematic, but instead an illustration of geography's power. He both identified the terrestrial physiognomy as geography's central object and implied that it was a legitimate undertaking for geography to examine the influence which the physical world had on human societies, apparently seeing no contradiction between them. In fact, rather than backing off the domain staked out by the social morphologists, he argued that in France one must not forget the usefulness of linking the teaching of geography to that of "societies" and that one ought to look to the American example where some strove for a "rapprochement between geography and sociology or history."[34] As examples, he cited Ellen Churchill Semple and Albert Perry Brigham, both known for using physical features to explain American history, precisely the sort of explanation that the Durkheimians attacked.

While the sociologists were busy attacking both history and geography, Vidal consolidated his position through his contribution to Lavisse's *Histoire de France depuis les origines jusqu'à la Révolution* and through the supervision of a number of doctoral dissertations. Both *normaliens*, Dreyfusards and promoters of the Republic, Lavisse and Vidal had had similar career paths; Lavisse was a member of the ENS *promotion* just a year before Vidal's and was also appointed to the ENS just a year before him. In his *Tableau de la géographie de la France* (1903) which introduced Lavisse's series, Vidal succeeded in making a case (albeit an unclear one) for an independent science of geography.

Vidal's volume was to serve as an introduction to what was to be a major collective history of France, ostensibly playing geography's traditional role as an auxiliary of history, and as a result, he was somewhat limited in what he could do to show geography's abilities. Much earlier precedents for such an introduction could be found in the geographical introductions to histories of France by Jules Michelet (1833) and by Victor Duruy (1860), both of which were clearly works of historians.[35] However, the length of Vidal's work, which surpassed by far the other two, enabled him to accomplish two goals. Not only did he meet the needs of the historians, providing them with a description of the various regions of France, but he was also able to demonstrate the growing stature of geography.

In part this dual purpose was satisfied by dividing his work into two parts. The first, entitled "Geographic personality of France," examined "the general relationships" contributing to the character of France and allowed Vidal to elaborate on his themes of the importance of position and of routes of communication. There he argued that France was characterized by its physical diversity but that due to a happy mixture of elements and to the crucial routes, it formed a "geographical being." Even in the second part, which was devoted to regional description, Vidal's work contrasted with those of Michelet and Duruy in its detail, in the extent to which he drew on the findings of the physical sciences (especially geology), and in his attempt to provide causal explanations.[36] Vidal's use of physical science remained, however, somewhat superficial. As Jean-Claude Bonnefort has noted regarding the section on Lorraine, Vidal's discussions of hydrographic basins and glaciation were dated and his description of the climate was rather vague given the information then available. For all its scientific trappings, Vidal's work remained impressionistic.[37]

Despite indications that Vidal saw some dangers in the explanation of social phenomena by physical ones, he did in his *Tableau* attempt such explanations. He noted the importance of human intervention in his discussion of routes, but he ended by stressing the importance of physical causes for history. This approach probably stemmed in part from the organization of the series. His work was to serve as an introduction to a history that stopped at the French Revolution, before many of the major transformations of the economy and landscape in which human intervention was so obvious.[38] In addition, his own discomfort with the increasing centralization within France may well have contributed to his rather deterministic interpretation. He seemed to long for a revival of local life,[39] within a unified France – something he felt could be accomplished by a return to the values of the "soil" and an appreciation of the contributions made by France's physical diversities. Sympathetic to regionalist sentiment, Vidal promoted not separation but a more egalitarian nation that took into account the needs and sentiments of the regions.[40]

In spite of the rather mixed character of Vidal's book, it was hailed as a masterpiece. Sympathetic geographers were quick to claim that this work demonstrated the great distance their field had come under Vidal's leadership;

they pointed to it as a methodological model. Bertrand Auerbach, professor of geography at Nancy, claimed that Vidal's work demonstrated how in contrast to "this somewhat bastard science which calls itself historical geography," *la géographie humaine* was established as a science.[41] Adulation also came from outside the discipline. In 1906, the Académie des Sciences Morales et Politiques awarded Vidal "le Prix Audiffred"; this was the first time the entire prize (worth 5,000 francs) went to one book. Speaking for the commission which awarded that prize, Gabriel Monod pointed to Vidal's "Rigorous scientific precision" and claimed that his work was far more successful than the works of Michelet and Duruy.[42]

The response from the Durkheimians was less clear since the review of Vidal's *Tableau* in the *Année Sociologique* was written by Antoine Vacher. As one of Vidal's doctoral students, Vacher was in no position to write a scathing critique, particularly since he was at this time somewhat at odds with Vidal over the character of his thesis.[43] Even so his review did indicate some reservations. He noted, for example, the incredible breadth of knowledge needed to succeed in such an ambitious undertaking. He also hinted that the work would soon be dated given contemporary research in geology and climatology and stressed that "man" was capable of modifying "the laws to which beings and things are subjected." Furthermore, rather than pointing to his supervisor's book as one which demonstrated that geography was an autonomous science, he described the relationship between geography and history in much less flattering terms, writing, "geography furnishes a canvas; history embroiders a design on it."[44]

Perhaps even more important than Vidal's *Tableau* for the institutional success of *géographie humaine* was a series of doctoral dissertations. Completed between 1905 and 1909, these theses were written under Vidal's supervision, something which was only possible following his appointment to the Sorbonne in 1898. Each was devoted to a specific region of France and examined both local circumstances and general causes. These *thèses d'état* were, in short, to demonstrate that Vidalian geography was a vital science. Differing significantly in character from work in historical geography, history, and Durkheimian sociology, the monographs both served as basic texts for the new discipline of *géographie humaine* and helped it to expand its institutional basis. Four of the authors gained positions in the Faculties of Letters by the year following their doctoral defences (Demangeon, 1905, Lille; Vacher, 1906, Rennes; Blanchard, 1907, Grenoble; Sion, 1910, Montpellier).

The precise nature of Vidal's supervision is rather difficult to establish.[45] He did, it seems, insist that the regional monographs include at least some discussion of those topics which he identified with *la géographie humaine*. As Vacher wrote to Albert Demangeon following his successful defence, "If I included several chapters of human geography and of cartography, the fault lies with Vidal, who has always counselled me against a thesis of pure physical geography, as I would have liked to do."[46] On the other hand, Vidal did tolerate quite a variety in the contents and points of view of the theses which he supervised. Even for students whose orientations he did not entirely endorse or understand, such as Vacher and Jules Sion (who he felt was really a sociologist), Vidal played the role of "patron" effectively, supporting them at their defences and in career advancement.[47]

Although the "Vidalian" monographs each dealt with a specific region within or straddling the French border and included at least some discussion of physical and human phenomena, they differed considerably in their content and approach, as can be seen in their treatment of regional definition, *enchaînement*, social phenomena, and historical sources. Nevertheless, they each bore a Vidalian stamp.[48] Not surprisingly, all the monographs paid considerable attention to regional definition and description. It was the scientific discovery of new regions, after all, which helped to legitimize this new science.[49] They attempted to describe the boundaries as precisely as possible, usually implied relative stability of the significant boundaries, stressed the difficulty of regional definition, and often used the notion of zones of transition. There was, however, little consistency in their use of labels. Only Blanchard called his study area a "natural region," Vidal's term for the new geographical regions. Despite the variety of labels, most did follow Vidal in including both physical and human criteria in the regional definition, usually with an emphasis on the geology and the soil. (Gallois, who worked more closely with the students, had argued that natural regions should be identified solely on the basis of their physical characteristics.)[50] Only Vacher and Sion stressed social criteria, with Vacher claiming that Berry continued as a region mainly for reasons of tradition and Sion emphasizing the importance of *règles juridiques* for the unity of Normandie Orientale. In addition, the monographs often referred to the *pays*, areas that were locally named and identified which lay within their region, a consideration promoted by both Vidal and Gallois.

All the monographs also discussed relationships between physical and human phenomena in some form or another, but only Demangeon and Raoul de Félice couched their studies in terms of *enchaînement*.[51] Demangeon and Sion described relations between man and his environment as reciprocal. Vacher and Raoul Blanchard stressed man's impact, and de Félice and Camille Vallaux appeared most interested in the influence of the environment without, however, falling into the trap of simple environmental determinism.[52]

Treatments of social phenomena in the monographs ranged from the racial characterizations of de Félice, Vallaux, and Blanchard to the more "scientific" treatments of Demangeon, Vacher, and Sion. Although only de Félice indulged in extensive racial stereotyping, de Félice, Vallaux and Blanchard all adopted what could be termed a social reform attitude toward the "races" which they studied. They suggested that the difficult straits in which these people found themselves resulted from weaknesses in character. Demangeon's focus was on relationships between man – in the singular – and the land, so that instead of examining social groups carefully, he discussed settlement distributions and the like. Vacher's study was essentially a work of physical geography and as such discussed neither social classes nor class relations. Nevertheless, some indication of a sociological perspective can be found in his discussions of the force of tradition as related to an understanding of "le Berry" and in a reference to language as a social phenomena.[53] The most sociological of all the monographs was that by Sion. He paid considerable attention to social classes and their changing relationships as affected by changing class interests. Those interests were often tied to property rights, which he labelled as *faits sociaux*. Showing some sensitivity to the *mentalité* of such groups, he also examined evidence for the "collective consciousness."[54]

These monographs were clearly quite different from the monograph on the social morphology of the Eskimo by Mauss and Beuchat. That monograph differed from the Vidalian ones by its stress on factors of density and seasonal variations over those of land and by its focus on the particular character of the social life. Also significant was the insistence that even in such an extreme environment as the Arctic, social traditions could significantly affect settlement. Nevertheless, in their discussion of the effects of seasonal variations, Mauss and Beuchat appeared to give an even stronger role to environmental factors than many of the geographers, attempting to explain basic categories of thought by seasonal variations. This explanation was tempered, however, by their argument that those variations had at least in part a social basis. In addition, unlike the geographers, they attempted to uncover a general law of social life.

Although all the Vidalian monographs included historical information, they were quite different from the bulk of the work in history which still focussed on political history and great events and which was also concerned with a critical examination of sources. The historical information that was included in the Vidalian monographs tended to come from contemporary history (dating from the Revolution). Of all the monographs, Sion's was the most historical. He divided his discussion into historical subsections and was the only one to adopt a skeptical attitude toward the work of the *sociétés savantes*, not unlike that held by many professional historians.[55]

The range of approaches in the monographs was a demonstration that the new practitioners of *la géographie humaine* were far from sharing a common position on such basic issues as the terminology and criteria for regional definition and whether an examination of the relationship between societies and their environment should attempt an explanation of either one or both. This confusion is not very surprising given Vidal's own ambiguity and the lack of agreement between Vidal and Gallois. Nevertheless, though the tenets of their new field were far from clear, the monographs were clearly not works of history or sociology. They also differed considerably from the theses in *la géographie historique* which preceded them as they adopted a regional focus, paid far more attention to the contemporary period, and examined the relationships between man and his environment.

Given the distinctiveness of the Vidalian monographs, had they succeeded in establishing the basis of a new autonomous field? Although one reviewer, Lucien Febvre, seemed ready to grant them this new status, Simiand, speaking for the sociologists, had strong doubts. Not surprisingly, given Berr's interest in synthesis and interdisciplinary communication, the monographs were reviewed in the *Revue de Synthèse Historique*.[56] For this task Berr chose Lucien Febvre, a young historian who had already contributed to Berr's series "les Régions de la France." In addition to his contacts with Vidal and Gallois,[57] Febvre was a very close friend of Sion. They had been in the same *promotion* at the ENS (1899) and the Fondation Thiers (1903) and had traveled together in the years before the First World War.[58] Febvre would have known all the others, with the exception of Vallaux, from his time at the ENS.[59] In short by his training, interest, and associations, Febvre was apt to be fairly well disposed toward these monographs.

When Demangeon's thesis was published in 1905, Febvre was still at the Fondation Thiers, had only recently begun his collaboration with Berr, and so was not in a position to review the work. Later, however, he reminisced: "When I think of Demangeon, I am gladly carried back to the thesis defence in 1905 which brought us together – the young of the time – around an older student who seemed that day to carry the common flag."[60] Despite his favorable predispositions, Febvre did not review the other monographs with unqualified praise. For example, he criticized those of Blanchard and Vallaux for failing to depict geographical causes clearly. In addition, he attacked their introductory sections on physical geography for not being clearly linked to the body of their work.

In his reviews of de Félice and Vacher, Febvre clarified his views on the character of a good regional monograph. De Félice's work was said to present a particularly dangerous conception of geography as it strove to "say everything" and "be complete."[61] Febvre feared that such a conception of the field would reduce geographers to compilers, lacking their own goals, conceptions, and methods. Revealingly, however, Febvre observed that despite its conception, the monograph did give the reader quite a good sense of the region and thus stopped short of attacking *le plan à tiroirs*[62] as it has come to be known. In his review of Vacher's monograph, he charged that that work should not have been presented as a regional monograph since it was really a study of hydrology, something Vacher probably would not have disputed.[63] In brief, Febvre seemed to endorse Vidal's statements that regional monographs should be concerned with *enchaînement* and by implication with some form of explanation rather than simply a compilation of information or an examination of a narrow range of phenomena.

Not surprisingly, given their close friendship, Febvre's review of Sion was extremely complimentary, concluding that his monograph was "the perfect model."[64] He argued that it demonstrated a consistent use of a geographical perspective, concentrating on geographical causes while at the same time exploring alternative explanations. This review was, no doubt, exactly what Sion wanted as he complained to Demangeon that de Martonne and Vidal had portrayed his work as more sociological and historical than geographical.[65]

Whereas Febvre's reviews indicated that Vidalian geography was a promising albeit not yet perfected field, Simiand challenged its entire basis suggesting that only physical geography made any sense. In 1910, he published a scathing review of five of the monographs in the *Année Sociologique* attacking their central object, the region, and thus implicitly challenging the very basis of the new field.[66] In part his criticisms echoed those he had addressed earlier to the historians. By restricting their monographs to only a single small region, the geographers, he argued, prevented themselves from deriving any conclusive findings, which could only come from comparative study. As he remarked, "in such a complex matter, to limit oneself to a single observation is to condemn oneself, ahead of time, to be unable to prove anything."[67] Given the complexity of their subject matter, the researchers would be better advised to limit themselves to specific questions for a much larger area using the comparative method. They might examine, for example, forms of *habitation*, distribution of houses, or localization of industry as would be done by social morphologists.

Despite the claims made that the monographs were to lead to a scientific renewal of geography, Simiand found little agreement on their geographical content. The only characteristic which the diverse phenomena studied seemed to share was that of location, but that, as Simiand reiterated, was a characteristic of other phenomena not included. If instead the authors meant by geographical those phenomena for which location is an essential element, they should have restricted themselves to a study of the physical phenomena of the surface of the earth. However, not only did they study human phenomena, they often stressed human impacts on nature instead of vice versa and used human factors such as overpopulation to explain the character of their region, a form of explanation outside of their domain and competence.

For examples of their lack of understanding of such phenomena, Simiand, as in the "1903 debate," singled out the treatment of economic factors, which he said the geographers tended to reduce to technical factors. First of all, contrary to the geographers' implications, economic phenomena were often distinct and even independent of "technical conditions," and secondly, physical factors could not explain technical ones, e.g. the presence of a water course could not explain the presence of a water mill. The physical factors were mere conditions; by contrast the explanatory factors were human or psychological. Even for "phenomena of habitat" which appeared to depend so heavily on the physical conditions, the geographers could not provide an adequate explanation, and in fact often obtained conflicting results.

Vidal never responded directly to Simiand's attack. The following year he published a two-part article on the concept of *genres de vie* in which he expanded on this largely human but ill-defined factor. In this essay, he failed to give a very clear explanation as to why the study of *genres de vie* was geographical and in fact, even concluded that *genres de vie* were not only "powerful geographical factors" but also "agents of human formation."[68] As for the geographical content, he indicated alternatively that a *genre de vie* had a strong effect on the physiognomy of a country, that such effects acted through the intermediaries of plants and animals, that one could locate the original "core" of certain *genres de vie*, that a *genre de vie* just like a plant required a favorable space to expand, and that the *genres de vie* acted in competition and responded to changing transportation techniques. In practice, when speaking of *genres de vie*, Vidal tended to stress the material manifestations of culture, especially those he could portray as stemming from the nature of the soil, i.e. tools, food sources, agricultural methods, dwellings, transportation, and so forth.[69]

In 1912, Vidal published an article in the *Revue Bleue*, "On the meaning and object of human geography." There, not only did he fail to address the question of the relations of human geography and sociology, he did not even give a very clear definition of *la géographie humaine*. His most explicit statements were that "it brings a new and broader conception of the relationships between earth and man" and that it represents a more "synthetic knowledge of the terrestrial laws and the relationships between beings." Even these tentative statements were quickly qualified with the disclaimer that to "understand its meaning and scope" one must view it in relationship with the other fields with which it has affinities, i.e. since it was clearly not identical to those fields, it was justified as a discipline in its own right.[70] Such an argument would obviously not appease Simiand who had argued that the field had no logical basis.

What one did find in this article was a retreat into a scientistic view of the field as Vidal strove to link it with biological and physical science.[71] He noted parallels between human geography and ecology and drew on such metaphors as "human alluviums" in reference to groups of immigrants. In his discussion of man as a geographical factor, a question which he claimed was at the "heart of human geography," man was treated in the abstract form of the singular, rather than as a member of society, and his role in marshaling natural forces to his ends was stressed. References to the influence of the environment on man also demonstrated little acknowledgment of social forces, which were merely lumped into a milieu whose physical and biological character had been stressed. People were seen as little different from plants and there was none of Vidal's earlier references to the importance of social constraints, tradition, and the like.[72]

In two "pedagogical lectures" (one as part of a series at the ENS in 1913 and the other at the University of Paris in 1914), Vidal addressed the question of the distinctiveness of geography, drawing on any number of different characterizations of the field. For example, he noted that whereas several related fields were also inspired by the notion of "terrestial unity," geography was particularly concerned with a study of "realities"; it drew on direct observation rather than abstraction. In other characterizations he argued that geography studied the surface of the earth, that it was concerned with a study of "the force of the milieu" and adaptation, and that it was a descriptive science which, without renouncing explanation, had as one of its main tasks to locate phenomena.[73]

As before, rather than giving a specific definition of geography or engaging in a debate over its logical underpinnings, Vidal was content to show that it was distinctive. Comparing geography to history, he argued, "Geography is the science of places and not that of men."[74] However, he then argued that geography was concerned with a study of men "whether they were subject to the influence (of places) or whether they in turn modify the appearance (of those places)." In a somewhat different characterization he also argued "the notion of place represents for the geographer what the notion of time is for the historian."[75] In yet another characterization, he observed that the time frame of geography was both longer and more focussed on the contemporary period than that associated with history.

Despite Simiand's harsh criticisms, Vidal failed to make specific comparisons to sociology; sociology still, after all, lacked a strong institutional base. He did, however, call geography a natural rather than a social science and remarked, "geography is called upon to use the same sources of facts as geology, physics, the natural sciences, and, in certain respects, the sociological sciences."[76] He still believed that geographers should study social phenomena, praising the work of Demangeon, Sion, and André Siegfried, but failed to explain whether their treatments were geographical because they escaped generalizations, dealt with location, took the form of regional monographs, or were related to the concept of *genres de vie*.[77]

For his part, Durkheim kept his distance in these discussions. In 1913, he gave a critical review of Jean Brunhes' *La Géographie Humaine* in which, however, he came close to admitting the legitimacy of *la géographie humaine* as a science. Brunhes was criticized for his artificial and scholastic categorization of geographical phenomena into six groups of "typical phenomena" and for limiting his field to "material works" so that an examination of the human group was pushed into the background. Although expressing his continued preference for the term "social morphology," Durkheim suggested that if one wished to keep the term *géographie humaine* it must not lead one to do violence to the natural relationships of the phenomena.[78]

In this same issue of the *Année Sociologique*, Demangeon in reviews of

Semple and Vallaux appeared to endorse some of Durkheim's positions. Demangeon preceded Simiand at the ENS by one year and was now established as a leader of *la géographie humaine*. He argued for more careful examination of "human establishments," such as forms of habitation, and also observed that Semple's *Influences of Geographic Environment* overemphasized the factor of territorial size at the expense of "other factors of the life of peoples such as the population density and the state of civilization." Thus, he advocated an examination of two factors (*habitation* and population density) which Durkheim had charged that geographers tended to ignore. In addition he complained of "an abuse of geographical determinism."[79] Demangeon's review of Vallaux was not quite so harsh, but even there he charged that the work was too hasty, lacking documentation and thought.[80]

As in the case of the debate between the sociologists and historians, little was resolved in these discussions on the nature of geography and of social morphology. However, in contrast to that debate, the positions held became less clear with time. Vidal, after showing some signs of recognizing the importance of social constraints, retreated into biological and geological metaphors and seemed to grasp at any possible defence. Despite the increasing obscurity of Vidal's position, Durkheim appears to have softened his stance, perhaps content to let Simiand make the most direct attack while continuing to try to involve some of the younger geographers in his cause. Although important criticisms of geography had been disclosed in these discussions, the absence of direct confrontations between the parties meant that they were easier for the geographers to overlook as they got on with their careers in a discipline which now seemed secure.

From Marc Bloch one finds no statement comparable to his "Historical Methodology" notebook of 1906 with which to demarcate his very early responses to Vidalian geography. That notebook, after all, appeared four years before Simiand's attack on the regional monographs. In it, the closest he came to discussing geography was in his section on economic laws in which he related settlement patterns to the wealth of the area and to the form of property, rather than just to water sources. For more direct reactions to the issues of regional definition, the regional method, and the appropriate object of place or regionally oriented studies, one must sift through his early publications dating from his years at the Fondation Thiers. There he began to respond more explicitly to the questions raised by the increasingly ill-defined field of Vidalian geography and its challenger "social morphology." [81]

5

From the Fondation Thiers to the doctorate
Marc Bloch's emerging perspective

After his year in Germany, Marc Bloch spent three years at the Fondation Thiers, then two as a lycée professor, and almost five in uniform. His writings from this decade reflect both his predoctoral status with the Université and his reactions to the on-going debates over the relative merits of history, geography, and sociology. His developing critical stance toward *géographie humaine* and his increasing attraction to Durkheimian thought can be detected in a monograph on "l'Ile-de-France," his dissertation, and a number of articles and reviews. Despite his interest in the new methods, the approach which he took in his published work remained very much like that of his teachers – a reflection of not only his junior status but also of the disruptions of war. He focussed on a careful interpretation of documents related to the history of France and made only limited use of the concepts and methods promoted by the Durkheimians.

In 1909 Bloch entered the Fondation Thiers, which had been established in 1893 as a residence in Paris for a very select group of doctoral students following their successful completion of the *licence* (an academic certificate acquired a few years after the *baccalauréat*) or an equivalent diploma. Normally around five students were admitted each year for a maximum stay of three years. *Normaliens* were typically very well represented at the Fondation, and in Bloch's *promotion* three out of five were ENS alumni. In addition to accommodation in a mansion constructed for this purpose, students received various forms of financial support including monthly allowances.[1] According to Emile Boutroux, a philosopher who was the Director of the Fondation Thiers during this period, the Fondation did much more than simply provide a studious atmosphere and the time and support needed to complete serious work. As he wrote, "there one enjoys the company of young men, no less excited by study, but devoted to works of a different sort, brought up in different milieux, following different methods; from this results an exchange of ideas which forms the judgment, and fertilizes the intelligence."[2]

For the most part, the range of topics and approaches of Bloch's contemporaries and those of the two preceding *promotions* (whose tenures would have also overlapped his) were in tune with the values of the Nouvelle Sorbonne. Three students (Davy, Gernet, and Granet) had established links with the growing group of Durkheimians and would pursue topics influenced by this association. Georges Davy had worked with Durkheim and Lévy-Bruhl and would also study with Mauss and Hubert at the Ecole Pratique des Hautes Etudes. Louis Gernet and Marcel Granet had both been involved with Lucien Herr's group of socialists,[3] which included many of the Durkheimians, and Gernet had contributed to Simiand's *Notes Critiques: Sciences Sociales.* Social and economic history attracted three *pensionnaires* (Georges de Lacoste, Henri Legras, and René Lote) and two others (André Tibal and Henri Alline) also demonstrated some interest in a social approach. The spirit of the Nouvelle Sorbonne was also evident in the attention paid to methodology. Ernest La Senne proposed to study whether logic could be constituted as a science on the model of physical science; François Gebelin, a *chartiste,* was preparing a "Parisian iconography"; and Alline drew on a method of "verbal statistics" in a study of Plato.[4]

In contrast to the *promotions* of 1907, 1908, and 1909, those of 1910 and 1911 demonstrated a resurgence of interest in letters with less of an attempt to relate developments to the social history of the times. (Examples included the work of Jean Ducros, Emile Pons, and André François-Poncet.) There were no new students tied as closely to the Durkheimians as Davy, Gernet, and Granet had been, although one student (Jules Pascal) did choose topics of social history, looking at property ownership in eighteenth-century France. In addition, there were two physical scientists, two in philosophy, one legal scholar, and one historian (Roger Doucet) interested primarily in political history. The reaction against the Nouvelle Sorbonne represented by the "generation of 1905" appears to have already set in.[5]

Although trends are difficult to assess when dealing with such small class sizes (varying from three to seven), it appears that, as had been the case at the ENS, Bloch's cohort marked a shift toward Durkheimian thought (at least for the *promotions* of 1907 to 1909) and a distancing from that of Vidal. It is understandable that the attraction toward Durkheimian sociology was particularly high during Bloch's tenure given that these men received the bulk of their university education immediately following Durkheim's arrival in Paris. Also, whereas there had been two geographers (René Musset and Jules Sion) associated with Vidal during the preceding five years, there were no students preparing geography theses during the five *promotions* which overlapped with Bloch's (though Pascal did adopt a regional approach in his study of the disappearance of the "anciens droits d'usage").[6]

During Marc Bloch's stay at the Fondation, his interests shifted from social and economic history to broader sociological questions, although still treated from an historical perspective.[7] Bloch's early interest in social and economic history is evident in a letter of application to the Fondation, written before his year's study in Germany. Describing his research topic as "the disappearance of rural serfdom in the Parisian region, and . . . the social, economic, and juridical phenomena which accompany it," he proposed to examine the dates of that disappearance, the character of the negotiations between the serfs and lords, and the legal condition of the serfs before and after their emancipation. In addition he would examine such social phenomena as a substitution of "limited leases" (*baux à temps*) for "perpetual tenures," a payment in money for payment in kind (particularly as related to Bücher's theories of a medieval economic revolution), and migrations between villages. Bloch elaborated, "My desire would thus be to study the date and the nature of these diverse social transformations, and to establish their mutual relationships." A final proposal was to study texts, particularly those of preachers, which could shed some light on the changing ideas which may have influenced or reflected the social transformation. This part of the research, Bloch explained, he had yet to start.[8]

For this study, Bloch proposed a regional framework. He explained that a study confined within geographical boundaries could be more useful than the accepted approach of studying a large seigneurie whose lands were scattered across various regions. However, instead of identifying what the Vidalians would consider to be geographical boundaries, he suggested using the Diocese since it was the only medieval territorial division for which boundaries could be established with precision. Bloch intended to complement his studies with comparisons to other French regions such as Normandy, where serfdom disappeared earlier, and Burgundy, where it disappeared later, and even to draw on foreign examples. He also noted that work done in Germany and England on the emancipation of serfs and the accompanying social transformation could offer useful insights, particularly given the absence of any similar work for a French region.

During his first year at the Fondation Thiers, Bloch spent part of his time selecting the region to be studied. After preparing two maps on serfdom and its disappearance, he chose a region that, according to Director Boutroux, was close to that known as Ile-de-France. In addition, he examined the institution of serfdom. He noted both the fluid nature of this concept in customary law and the conflicting links of the concept of "serfdom" to both ancient slavery and medieval vassalage. By Bloch's second year, Boutroux mentioned Bloch's interest in the transformation of serfdom from that of a personal bond and reciprocal obligation to that of a serf as a member of an inferior social class, and also his study of the significance of the acts of the French kings in emancipating their serfs. The latter was to be developed as Bloch's complementary thesis with the title "La politique financière des derniers Capétiens et les populations serviles." By the end of Bloch's third year, his work was described as the evolution of the social system of serfdom. He found that economic factors were less important than often suggested but that religious ideas could be viewed as significant "agents of transformation of society and even of the economic conditions themselves."[9]

It is not possible to establish the mutual influences of Davy, Gernet, Granet, and Bloch, but it does not seem unlikely that Bloch's increasing sociological thrust was at least partially a reflection of his association with these three colleagues. Ricardo Di Donato has argued that Gernet, Granet, and Bloch formed a research group at the Fondation and that Granet had a particularly strong influence on the other two.[10] More direct testimony comes from Davy, who noted the close working relationship between Bloch and Gernet at the Fondation Thiers.[11] Henri Lévy-Bruhl, another member of Herr's group of socialists, suggested that Gernet was "a great friend of Marcel Granet."[12] Correspondence between Bloch and Davy indicates that Davy was already a close friend of Bloch's at the time of their applications to the Fondation Thiers, a friendship which they maintained in later years.[13]

Although Bloch had intended to examine religious ideas even before entering the Fondation, his increasing emphasis on the transformation of the institution of serfdom and his down-playing of economic factors in favor of religious ones was entirely compatible with the types of research done by his three colleagues. They had stressed religious rites and the broader societal transformations over economic explanations. Gernet, for example, had treated economic phenomena from a sociological perspective as in his *mémoire de licence* (1903) in which he stressed the importance of "social psychology" for an understanding of the origin of money. Similarly in a 1909 paper on the provisioning of Athens with wheat, he identified his object of study not as the commerce itself, but instead as "the collective psychology of the Athenians." This study was praised by Simiand for its attempt to explain the economy in terms of states of collective psychology rather than vice versa.[14] For his part, Granet, who first proposed a study of "the sentiment of honor in feudal societies," later focussed on the religious rites which were at "the foundation of Chinese society." Davy, in turn, sought to explain the evolution of contracts through an examination of religious rites such as the potlatch.[15] Bloch's interests in changing social obligations and bonds, the legal interpretations of those bonds, and comparative research pre-dated his years at the Fondation Thiers; nevertheless it does seems likely that his close association with Davy, Gernet, and Granet at the Fondation helped to bring him even closer to a sociological perspective.

By contrast, Bloch did not encounter any strong representatives of Vidalian geography at the Fondation. Though he intended to use a regional framework, he attempted to define that region by the social criteria of serfdom and its disappearance rather than by geographical ones. In general, the nature of Bloch's research was well within the lines of that done at the

Fondation during his tenure, particularly that of his class and the two preceding ones. It differed, however, from the tone for the two *promotions* which followed his. In these the social interests were represented only by Pascal whose approach was closer to that practiced by Bloch before entering the Fondation than that which he espoused following his years as a *pensionnaire,* i.e. that of a social and economic history which only drew to a limited extent on geography.

Bloch's first major publication was his monograph on the Ile-de-France, the last in Berr's series "Les Régions de la France," which was published in the *Revue de Synthèse Historique* from 1903 to 1913. In his statement of intent for the series, Berr described the monographs as studies into "the psychology of historical groups." He felt that a study of groups at the scale of "a people" was at that point too vast and vague, as demonstrated by work on *Völkerpsychologie* in Germany, and that a study of "race" was also inappropriate. Instead, Berr advocated an investigation of "historical individualities" of smaller size which in the series would be attempted on a regional basis. He hoped that the monographs would examine well-defined groups, drawing on information from geography, history, folklore, literature, art, and religion, and attempt to discover how they had been formed, what they had experienced, and whether they continued to survive as a group. As for the regional definition, Berr explained, "We are not preoccupied with making a prior division – by natural or historical regions. That which could seem rational would have been in reality, arbitrary. It is up to each of our collaborators, through his work, to justify the determination of the subject."[16]

Berr's only other major stipulation was to establish a format for presenting the authors' findings. Each was to evaluate the state of the work done on the subject, to summarize the findings of that work, and to discuss in the form of hypotheses and partial syntheses what remained to be done. The professionals were to lead *les savants* onto the right path, presenting their findings in a pedagogical format.

With his aim of bringing disciplines together in a rational synthesis, Berr foresaw drawing contributors from both geography and history and in his initial statement said that both Vacher and Demangeon would contribute, although in the end neither did.[17] Instead, many of the contributors were drawn from precisely those in whom the Société d'Histoire Moderne had placed their strongest hopes for a reorganization of local history, those holding chairs of regional history in the provincial universities. In addition, two were associated with libraries, and two (Bloch and Febvre) were *pensionnaires* at the Fondation Thiers.[18]

Of all the monographs published in this series, Bloch's was the farthest from the common mold. Like the others he followed Berr's pedagogical format quite closely, but unlike them, he was not as directly critical of the work produced by the local historians. Other contributors had described this local work as overly patriotic and polemical, characterized by a poor choice of subject matter, and as poorly documented and written. The *savants,* they charged, tended toward oratory and philosophizing rather than serious scholarship.[19] By contrast, though making many specific recommendations on research tools and themes, Bloch took a somewhat more diplomatic approach arguing that the professional historians owed the societies a debt of gratitude for the work they had accomplished. On the other hand, he did make a direct attack on the whole concept of local history. In his discussion of village monographs he argued: "Despite the zeal displayed by their authors, they are too often useless for general history, that is to say, when all is said and done, for the only history that matters."[20] Bloch concluded that "a good study of local history could, no doubt be defined as follows: a question of general interest posed to the documents furnished by a particular region."[21]

Despite Berr's designation of "the psychology of historical groups," as the subject and synthesizing theme of the monographs, most of the authors gave it only superficial treatment, seemingly content to recite ethnic stereotypes.[22] Discussions of "race" were also common despite Berr's caution over the use of that term. Variations recognized tended to be regional ones, with little attention given to changes in the composition and character of *individualités historiques* and only an occasional reference to social class. In the second monograph, Sebastien Charléty had shown rather more sensitivity to the problems with such an approach, making only tentative generalizations around such phrases as "common habits born from similar interests."[23] Bloch went even further. He stressed the lack of a coherent social group for his "region" and made frequent references to social class and to social, political, and religious movements whose base spread far beyond the Ile-de-France.[24]

Even though Berr gave the contributors a free choice in their definition of region, most chose what might be called an historical region, showing little attraction toward the natural region common in the geographical literature of the times.[25] One of the most succinct statements on this issue came from Henri Prentout: "Normandy is not a geographical entity; it is above all an historical expression. Like many other lands, this province is a work of history."[26] Among the criteria used to designate these regions were political, linguistic, ethnological, and even moral ones. A common method was to define a region by its center.[27]

Bloch was more critical of the regional approach than his colleagues. He argued that his study area was "lacking regional unity," be it geographical or historical, and directly challenged the usefulness of geographical concepts such as natural regions and *pays* for historical work. Not unlike some of Gallois' work,[28] Bloch began his monograph with an examination of the origins and changing meaning of "Ile-de-France" and concluded that what all the meanings had in common was simply the same center, Paris. Adopting a term coined by Vidal in the *Tableau de la Géographie de la France,* he referred to the area as simply, "the *pays* around Paris," but unlike Vidal he denied that this area had any regional unity. Furthermore, although noting that in certain parts of his study area, such as Beauce, one could identify a *pays* in the geographical sense, he argued, "In sum, the *pays* rarely offers a convenient framework for historical research."[29] Bloch's critical treatment of *pays* contrasted not only with the Vidalians but also with the monographs of Febvre and Christian Pfister (his thesis supervisor), both of which drew unquestioningly on the concept.[30] He did, however, admit that the terms of "popular geography" might occasionally have some use in locating very small areas defined by their agricultural characteristics. More generally, however, Bloch argued, "The limits of the field of observation must vary with the object which the scholar observes. . . . There are no ready-made regions with which the historian could rest content whatever he studies. Depending on the question addressed, he will make his own region which will be different each time."[31] Given his interests, he seemed to favor those regions which encompassed a coherent social group or in which one could accurately speak of a regional life. Such was not the case for Ile-de-France. For this reason, no history of Ile-de-France had been written and in Bloch's eyes never should be – in contrast to certain areas which had once been independent states (such as Franche-Comté, Lorraine, Brittany, and Normandy).

On the subject of studying particular places, Bloch argued that such studies must be designed to serve wider ends. In his words, "To write the history of a town is not to recount a sequence of phenomena with no other link between them than the poor link of the unity of place" – a clear contrast to Vidal's statement that geography could be defined as the study of places.[32] Monographs of towns could usefully draw on the townscape and town plans but only if related to social and economic phenomena. The key was to seek how its entire history was "translated" in the monuments, the plan of the town and its material history. On the other hand, Bloch discouraged monographs centering on a single village. There, the source material was apt to be insufficient and difficult to interpret, requiring an enormous breadth of knowledge, which most of those attracted to such topics lacked. Instead, he suggested that local historians would be better advised to research one aspect of that history across a wider area, a suggestion very much like that of Simiand in his review of the regional monographs. Examples given were a study of the churches of the *pays* and a study of the political life of the district at the time of the Revolution.[33]

Despite his reservations over some geographical techniques, Bloch implied that work in both historical and Vidalian geography could prove useful to the historian. For example, speaking of works in local history, he wrote, "Most often, one must admit, what we ask of them is not so much history itself, as the material of history: scholarly dissertations on details of history or historical geography, lists of bishops or abbots, genealogical tables, supporting documents."[34] One of the key spokesmen for traditional historical geography, Auguste Longnon, was praised for his work on Ile-de-France, which demonstrated, according to Bloch, that the boundaries of the *gouvernement* of Ile-de-France had been determined by history, not geography.[35] However, later on, when discussing "what one could call the geographical history of Ile-de-France," Bloch asked, "Which lands, in Ile-de-France, attracted the first human establishments? Valleys or plateaux? And amongst the latter which more so, the dry plateaux or the damp ones?"[36] These were the sorts of questions addressed by Vidalian geography, which attempted to relate physical features with human settlement. Also, like the Vidalians, he did not favor simple geographical determinism, arguing that it was a very delicate task to sort out the respective roles of geographical and historical factors.[37]

In part the distinctiveness of Bloch's monograph can be attributed to its appearance as the last of the series (appearing after Simiand's attack on the Vidalian regional monographs) and to the peculiar nature of the "region" which he studied. However, it also reflected his views of the character of history. Its rather tedious character, which has been attributed to Bloch's cautiousness and inexperience at this early stage,[38] stemmed in part from the confines within which he worked. In a piecemeal and indirect fashion imposed by Berr's structure, Bloch successfully attacked the concepts of region, historical group and local history, and praised the *sociétés savantes* – quite a departure from the initial intent and tone of the series.

Although Bloch's monograph did not fit the mold of Berr's series, it was closer in character to it than to the Vidalian monographs. Their authors had put much greater emphasis on regional definition using physical criteria as well as some human ones, and on reciprocal relations between man and the land. Going farther than the others in Berr's series in his criticisms of Vidalian techniques, Bloch challenged the concepts of region and *pays* and defended historical geography. Also like the other monographs for "Les Régions de la France" series, Bloch paid more attention than the Vidalians, with the possible exception of Sion, to history and historical methodology. The opposition between Bloch's monograph and those of the Vidalians was not, however, complete. For example, his avoidance of racial stereotyping had more in common with their reluctance to use such stereotypes than with the frequent recourse to such depictions in the monographs written for Berr.

Bloch's monograph, in short, fitted neither mold well and was not simply a synthesizing of the two approaches. It had already become influenced by another school of thought, that of Durkheimian sociology. In his criticisms of the regional approach and of the "unity of place" and in his treatment of social phenomena, Bloch had in fact adopted an approach not unlike that of Simiand in his review of the Vidalian monographs; he pushed for comparison, was suspicious of location as a unifying force, and focussed on social phenomena. Even though his first major publication had been a foray into local history and geography, Bloch clearly did not intend to continue along those lines.

Additional evidence of Bloch's early views on geography can be found in three reviews which he wrote for the *Revue de Synthèse Historique*. In the first Bloch criticized a work by Marcel Poète for neglecting "the geographical factor." The second was a short note on a posthumous publication by Auguste Longnon, which Bloch praised for its "clear" and "objective" treatment of the history of the ethnic formation of the French people and of the "territorial formation" of the French state.[39] The last review was a more extensive one of a book on the Franche-Comté that Febvre wrote for the series "Les Vieilles Provinces de France" (a series which relied heavily on the historians who had written monographs for Berr).[40]

Although Bloch and Febvre must have known of each other, given their collaboration with Berr, Febvre has testified that until they both took teaching posts at the University of Strasbourg after the war, they had only met once, at the Blochs' in 1902.[41] In his 1914 review, Bloch approved of Febvre's use of geography, singling out a chapter on *genres de vie* as of particular merit. Bloch wrote, "It will be necessary from now on to recommend this chapter, or even the entire book to whomever still doubts the utility of a solid geographical education for historians." He also liked Febvre's description of the region as "an assemblage of natural regions, broken, cut up, and united in a mostly political ensemble" – an ensemble which economic bonds had helped to shape. In a comment echoing those he made on the Ile-de-France, Bloch argued that a history of Franche-Comté was justified because it was "truly a province, endowed with an autonomous life, and at certain moments, almost an independent state."[42]

Though praising Febvre's use of geography, Bloch did have some reservations about Febvre's monograph. He questioned Febvre's emotional style which was, he claimed, closer to that of Michelet than that of Fustel de Coulanges. In addition, he suggested that the treatment of the history of Burgundian patriotism was inadequate, and that Febvre's use of collective psychology was dangerous. Febvre, he explained, had a tendency to attribute characteristics to *les Franc-Comtois* in general rather than to social groups such *les paysans* and *les petits-bourgeois*. It was, Bloch felt, far too early to make such generalizations. He cautioned that to neglect the established historical principles of prudence and methodological doubt would be to "compromise the future of collective psychology."[43] In brief, although endorsing Febvre's use of geographical concepts and approving of his interest in social history, Bloch was uneasy with Febvre's somewhat glib style and his treatment of social phenomena in *Histoire de Franche-Comté*. Febvre may have understood and come to terms with developments in the field of geography, but he did not have the same understanding which Bloch had of developments within the field of sociology.

Although the evidence is not extensive, Bloch does seem to have placed some value on the geographical training which he received.[44] He was ready to admit that a regional approach could occasionally be useful if done with due attention to historical factors. In addition, he argued that one could not totally neglect the "geographical factor," and he was quite taken with the concept of *genre de vie*. On the other hand, being outside of the Vidalian circle, he also seemed to find value in the work of *la géographie historique*. He had learned from the Vidalians, but remained very much an historian.

Following "L'Ile-de-France," Bloch's next major publication was his dissertation. This work, taken together with some of his other articles and reviews, can illustrate his developing positions on the historical and sociological approaches. As a result of his military service between 1914 and 1919, Bloch's plans for his doctorate were radically modified. Given the circumstances, he chose simply to expand what had been intended as his complementary thesis instead of writing the major work on the evolution of the social system of serfdom formulated during his stay at the Fondation Thiers – a decision which undoubtedly served to dampen the sociological character of his work. The title was changed to "Rois et serfs. Un chapitre d'histoire capétienne."[45] Originally designed as a criticism of two famous documents, the revised version attempted a broader examination of the emancipation of serfs on royal lands and included material on the development of "servile institutions," that would have been included in the principal thesis as first planned. The thesis focussed on economic themes and relied heavily on the rather traditional exercises in documentary criticism, which were more representative of his thinking and training on leaving the ENS than on his departure from the Fondation Thiers.

The main argument of his thesis was that the two famous *ordonnances* (a misnomer since they were actually "letters of commission") of Louis X and Philippe V were in no way exceptional. Contrary to the accepted interpretation, these documents did not indicate an emancipation of all the serfs on crown lands. Instead they were directed to clerks who were to go only to the prosperous *bailliages* of Senlis and Vermandois in an attempt to convince the serfs to buy their freedom. Such a practice was similar to emancipations carried out under the three preceding French kings – emancipations which did not blaze the way for those undertaken by the seigneurs and religious orders but instead were very much part of a more general trend.

Although the economic motives of the two famous emancipations had been acknowledged, historians had also assumed that they represented general emancipations citing as evidence the preambles to *les ordonnances*. Those preambles referred to the natural right by which all men are born equal, indicated that the word "France" was derived from *franc* meaning "free," and, even more notably, appeared to order that "generaument par tout nostre roaume de tant comme il puet touchier a nous et a nos successeurs teles servitudes soient ramenees a franchise."[46] In his documentary criticism, Bloch argued that such phrases were little more than literary conventions. They represented survivals from as early as Roman times written by functionaries steeped in notary conventions, who may well have had no direct knowledge of the specific intent of the letters of commission. Rather than indicating a radical departure, these phrases were simply hackneyed expressions.

In his heavy reliance on methods of documentary criticism, Bloch's approach was closer to that of the Ecole des Chartes than to the Durkheimians. Major portions of his thesis were devoted to a discussion of the character of the sources, and his central point was established by using the principle of verification by sources with "absolutely distinct origins."[47] Bloch did note emancipations in other regions of France. However, such comparisons served simply to show changes in royal policy rather than to compare a given institution in societies representing specific social types in order to uncover specific causes and laws, as Durkheim might have advocated.[48] No discussion of parallel developments for other countries was given, although he did refer to the German, English, and Italian examples in his discussion of literary conventions.

Similarly, Bloch's treatments of language and of economics were not particularly Durkheimian. In the discussion of literary conventions, his analysis of linguistic survivals was more one of literary history than of the linguistic analysis so enthusiastically adopted by his colleagues Gernet and Granet.[49] Economic causes rather than broader social change were stressed. Thus, Bloch noted how specific political events created economic needs which in turn led to the emancipations – rather than trying to understand an economy by states of collective psychology, as in Gernet's early work.[50] In his conclusion, Bloch stated succinctly, "The manumissions thus depended on economic conditions."[51]

Despite his reliance on the methods of political, literary, and economic history, Bloch's treatment of his thesis topic also indicated that it was not a work of the history establishment of which Simiand was so critical. He had managed to escape at least two of the three dreaded "idols." By putting two of the key documents of medieval history in a broader context and by revealing their exaggerated significance, he showed that the emancipations were certainly not the work of two wise kings, and thus he avoided the trap of the "individual idol." Furthermore, though noting political factors, Bloch's focus was on the broader process of emancipation and so was not strictly speaking an example of the "political idol." Finally, although the book relied on chronological explanation and narration and did not include a discussion of the "normal type," Bloch neither concentrated on a search for origins nor treated all periods as equally significant, so this was not a particularly good example of the "chronological idol" as described by Simiand.

It was in the incidental commentary and discussion rather than in the main argument that Bloch appeared closest to his sociological colleagues. For example in a discussion of *le chevage,* a small yearly tax, he demonstrated his interest in formalism and ritual:

In the 11th century and at the beginning of the 12th, the *chevage* was the characteristic burden of the servile condition: even better its symbol. To make oneself the serf of an abbey, or to recognize oneself as such, an oral or written avowal was not at all sufficient: there had to be a material act, a ceremony inspired by the formalist spirit of old medieval law.[52]

Bloch also noted that during this early period ties between a serf and his lord were strong and could only be broken by juridical process marking the rupture. He lamented that little was known about "exterior forms of manumission."[53] Later in his discussion of the limited documentary evidence for the tenth and eleventh centuries, he wrote that in those periods:

The majority of the juridical processes took place according to the forms of a symbolism which was self-sufficient; they took shape in the gestures performed according to rites, prescribed by a custom that was otherwise not particularly rigid; they proved themselves in the oral testimony. The written act was a sort of luxury, hence its rarity.[54]

This interest in formalism and symbolic rites paralleled discussions in the writings of Davy and Granet. Davy's thesis emphasized the "contractual formalism" which he associated not only with "potlatch societies" but also with old German law and with the situation when classical rights were first instituted in ancient Rome. Granet, in turn, was particularly interested in the social function of the seasonal rituals which established alliances through group marriages between different local groups.[55]

When discussing literary conventions, Bloch stressed the difference between medieval and contemporary thought. He referred to the "strange customs" which allowed an expression of charity and financial motives in a document freeing serfs and remarked that it was not unknown to include a statement freeing serfs in a will when it was clear that such an act would not take place. Such an assessment paralleled those of Gernet for ancient Greece and Davy for the Kwakiutl in noting how much early conceptions of law and morals differed markedly from those of the contemporary period.[56]

In contrast to Davy, Gernet, and Granet, Bloch did not try to document wide-ranging evolutionary changes from "primitive" to classical or feudal times or to extrapolate a wider significance for similar societies. He did, however, discuss changes in the character of the institution of serfdom, in the emancipations, and in the structure and character of the state. He was not totally immersed in the particular and was interested in tracing institutional change, albeit in a more restricted way (in both place and time) than the sociologists.

Although discussions relating to geography and social morphology in *Rois et Serfs* were limited, Bloch did originally intend to write a regional thesis. As he explained in the preface to *Rois et Serfs*:

> In fact at the same time, I intended to publish a broader work, as a principal thesis, which I planned to call: *Les populations rurales de l'Ile-de-France à l'époque de servage.* There I would have recounted, in a regional framework, the evolution of serfdom as I had come to picture it after rather long research.[57]

Significantly, the region was not to have been the object of study but instead simply "a framework." At this point, Bloch insisted that he planned to complete his interrupted project. That work, however, was never written, and one wonders whether Bloch's disenchantment with the regional approach, as demonstrated in his monograph on the Ile-de-France, was a factor. Bloch described the "essential object" of this proposed work not as the region of Ile-de-France but as the question: "What is serfdom?"[58] Bloch never did pursue his regional work on the Ile-de-France but he did return to the topic of serfdom several times, developing themes closer to those of the sociologists than to those of the geographers.

Despite his initial plans to use a regional framework, Bloch made no particular attempt to define or defend a regional approach in his dissertation. Neither did he explicitly adopt the approach of social morphology, even though he was sensitive to specific material conditions of the society which influenced the emancipations. The statement which came closest to an examination of the material form of the social reality, to borrow Davy's expression,[59] was the following:

> It is natural that the urban populations gained their liberty more quickly than the people of the *plat pays.* The economic importance and even more, without a doubt, the military importance of the towns encouraged the kings toward conciliation. Squeezed behind their confining walls, the bourgeois, more than the peasants, had the power and the taste for collective action. Finally and mostly, they were richer.[60]

However, even in this case, it was economic factors that he stressed.[61]

Bloch expressed his views on the sociological and historical approaches most clearly in his articles and reviews, rather than his longer monographs. The lengthy article entitled "Les Formes de la rupture de l'hommage dans l'ancien droit féodal," written at the Fondation Thiers, was the most sociological. Formalism and ritual were now the focus, not just an aside. Bloch attempted to search for the traces of "a formalist ceremony of the rupture of homage," arguing that such ceremonial forms were necessary to make the juridical processes obligatory.[62] Nevertheless, the article was much like *Rois et serfs* in terms of methodology. The major argument was again established by examining and comparing a variety of sources, and although Bloch examined the meaning of expressions, he did not engage in a detailed linguistic analysis. Also, although Bloch gave some attention to the decline of formalism, his major thrust was simply to establish the existence and meaning of this particular rite during which the bond of homage was terminated – the rite in which one party threw or broke a twig or wisp of straw in the presence of the other. Bloch's approach here contrasts with those of his sociological colleagues who studied rites and formalism in order to trace the transition from one form of society and thought to another, or to uncover a wider social significance amongst similar societies.

Bloch did demonstrate an interest in more general questions relating to the symbolism of the twig or stick and to the origins of feudalism but hesitated to draw broader generalizations. According to him, the stick that was broken or thrown was not a uniform symbol, but instead its meaning varied with the situation. Nevertheless, there were parallels to other ceremonies which Bloch found intriguing, for example, that by which, in the case of the Franks, one abandoned his family by breaking sticks over his head and tossing them in four directions. To establish a link between the act which broke the family bond, "the strongest of the social ties in old Germanic societies," and the act which broke the tie of vassalage, the cornerstone of a new society, would, he wrote, be of considerable interest.[63] However, given the lack of evidence Bloch felt little could be concluded. Though intrigued by the questions sociologists would ask and certainly not mired in the particular, Bloch was rather cautious in his conclusions, preferring to point to interesting questions rather than to establish definitive answers.[64]

Bloch's other early published writings dealt less directly with sociological topics and methods. Nevertheless, in a number of reviews he demonstrated his awareness of the orientations of the Durkheimians toward economic history. He appeared to have no trouble viewing economic history as a social science and appeared to find sociological concepts useful in such studies.[65]

Despite the insights which Bloch obviously found in the sociological works, he did not favor an overly theoretical approach. In 1912, he criticized Andreas Walther's book, *Geldwert in der Geschichte* for being too obscure and abstract and also faulted Walter Layton's *An Introduction to the Study of Prices* for giving a doctrinal statement before presenting the facts. Not unlike Seignobos, he argued that the goal of the historian was to uncover "the reality of the facts of history, not to engage in polemics." As he wrote in a 1918 review of Georg von Below's *Der Deutsche Staat des Mittelalters*, "the book seems only to be a long discussion, an incessant shock of theories."[66]

On the occasion of the distribution of prizes at the lycée in Amiens in 1914, Bloch gave a speech entitled "Critique Historique et Critique du Témoignage" which illustrated that he was not ready to make a radical departure from the principles set forth in the basic text by Langlois and Seignobos.[67] Like them, he stressed the care needed in historical reconstruction, the problem of conflicting testimonies, and the need to compare a variety of testimonies. The historian was to strive to establish the independence of the various testimonies and to assess the character of the knowledge held by witnesses. By such a method, he proclaimed, one could "extricate the truth." In this address, however, Bloch did expand the legal metaphor used by his predecessors who had described authors as witnesses. Bloch described the historian's role, instead, as that of a *juge d'instruction* (examining magistrate), a metaphor which he used to illustrate the very active role the historian must take in examining historical evidence.

The major contrast between Bloch's developing approach and that of Seignobos lay more in the questions which he favored than in his method of examining primary sources. Defending documentary criticism against the charge that it destroyed the poetry of the past, Bloch insisted that the poetry could still remain if only one used the sources for a different purpose. Langlois and Seignobos had cautioned against legends as a documentary source, explaining that they belonged to folklore, not history, and that they could only serve as a source of people's conceptions. For Bloch, however, that was precisely why they were of particular interest.

We treated them as bad chronicles. And here they are only beautiful tales! Now that we know how to read them, they offer a clear image: that of the heroic and puerile spirit, turbulent and eager for mysteries, from the century which saw them born. What constitutes the beauty of the legends and their own truth, is to faithfully translate the sentiments and beliefs of the past. To know them as legends, we appreciate them more.[68]

Even in his earlier writings, the questions which Bloch chose to ask were those of social and economic history rather than of a history of politics and prominent individuals.

In three early articles which were devoted to a critical examination of primary documents, Bloch, despite his focus, also indicated what he saw as the interest which the documents held. For example, he wrote in a bibliographic essay that the documents in question could help one to disclose the "functioning of the communal institutions of the burg of Chelles."[69] Bloch's article on Blanche de Castille, like his thesis, used documentary criticism to downplay the role of particular individuals in the emancipations stressing instead more broadly based forces. Similar interests were apparent in his early reviews. Thus C. L. R. Fletcher was criticized for emphasizing political history at the expense of the "history of institutions," ignoring the important question of the origins of feudalism. David Hill was attacked for reducing the history of the Middle Ages to "the struggle of the 'sacredoc' and the Empire" and glossing over important issues such as the relationships of the lord and the vassal, and Paul Viard faulted for his failure to highlight the economic importance of tithes.[70]

Another theme of interest to Bloch was that of nationalism and patriotism, but viewed as a social phenomenon rather than a political one.[71] Von Below was criticized in 1918 for exaggerating the role of the state in medieval Germany and for failing to address questions of patriotism. This, Bloch argued, was a common error of German political science "for whom the State is everything and the nation very little."[72] In a 1920 review of *Les Origines de l'ancienne France* by Jacques Flasch, he also argued, following Lot, that regional nationalities did not lead to a great number of small independent states in France following the dissolution of the Carolingian Empire. The great feudatories were tied to the king by homage, and regional nationalism was only present in certain areas at a later period when there were larger social unities such as the French, Burgundian, or Flemish *patries.* Although Bloch's thrust in these reviews was away from the state, individual personalities, and events, it was still more political than that with which some Durkheimians would have been comfortable.[73]

By the time Bloch completed his years of apprenticeship, his approach had already begun to diverge from that of the historical establishment. He continued to follow established historical methods and to share some of their concerns, but he also drew in different ways on both geography and sociology. He had written a regional monograph which attacked the concepts of "region" and *pays* and a history thesis which used accepted methods and yet introduced a few sociological concerns, shying away from the "idols" attacked by Simiand and touching upon questions of formalism. Showing some ambivalence toward geographical approaches, he noted the importance of "the geographical factor" in his reviews and praised the Vidalian concept of *genre de vie* despite his earlier criticism of some of the other Vidalian concepts and his praise for *la géographie historique*. In articles and reviews related to sociology and history, he examined rites and formalism (using established historical methods), indicated that economics should be viewed as a social study, but argued against an overly theoretical approach and advocated a close examination of "the reality of the phenomena of history." Also very much part of the thrust to create a new vision of "la France," Bloch focussed throughout on the historical center of France posing questions about nationalism and patriotism rather than past events and important individuals. Bloch still saw history as a search for truth, to be undertaken according to the rules of documentary criticism, but his vision of where the truth lay was a different one from that of his teachers.

PART II
Marc Bloch as a critic and practitioner of sociology and geography

6

The University of Strasbourg as a center of disciplinary change

Following the First World War, Marc Bloch was one of the first to be appointed to the reopened University of Strasbourg in repossessed Alsace. That university was reopened with high hopes that it could become both a model for educational reform in France and a showpiece for French culture. Both Alsatians and the world had to be convinced that France could do more with the University of Strasbourg than Germany had already done. It must become, in the words of the President of the Republic, "at the Eastern frontier, the intellectual beacon of France."[1] Once again Bloch was to be at the center of disciplinary change and development. During his years in Paris, he had witnessed the growth of Durkheimian sociology and Vidalian geography, and at Strasbourg he would be at the center of a disciplinary restructuring as efforts were made to overcome the relatively new disciplinary divisions. Not only did a number of the faculty work very actively at promoting interdisciplinary work but also the very structure of the university was designed to encourage innovative research and interdisciplinary exchange.

In mid-December 1918, a commission of seventeen prominent French scholars visited Strasbourg, just a few days after the closure of the Kaiser-Wilhelms-Universität Strassburg, to make recommendations on its imminent reopening as a French university. Gustave Lanson reported how they marveled at what the Germans had left behind. Nowhere in France, not even at the Sorbonne, were there equivalent facilities; the buildings, equipment, and number and salaries of the personnel all impressed the commission. Those members representing the various Faculties of Letters were particularly taken with the seminars, of which there were about sixteen. Each occupied two or three rooms and was provided with its own library and generous funding.[2]

Following their annexation of Alsace in 1870, the Germans had done their best to make the University of Strasbourg into a showpiece for the superiority of German culture at a time when no universities (in the modern sense of the word) existed in France. The German university enjoyed substantial subsidies from the Empire and massive appropriations for new facilities from the

Reichstag. It also acted as a center for educational reform. The faculty had a large proportion of young and promising scholars who participated actively in decisions on new appointments and in administration. Given their youth, professors were very open to new developments within their respective fields, and many were committed to teaching in seminars that were regarded as laboratories rather than teaching by lectures and oratory.[3]

Also significant was the university's excellent library. Having mistakenly destroyed the city library during their bombardment of Strasbourg, the Germans organized an extremely successful drive for a replacement. Contributions came from the emperor, around 300 publishing houses, and several foreign countries. By the early 1880s, the library had over half a million volumes making it the largest university library in the world, an honor it kept until the First World War when Harvard overtook it. The university soon established a reputation for both serious scholarship and modernity, and attracted top students from throughout Germany.[4]

Not willing to be outdone by the Germans, the French educators hoped to take advantage of the facilities at Strasbourg to build a French university that would surpass all other French provincial universities and compete with Paris. The University of Strasbourg, they argued, should be turned into a model university in which the vices of the faculty system could be overcome and a corporate spirit fostered. It was even suggested, by Charles Andler, that the faculties be replaced by institutes in which scholars could group themselves according to their research interests to encourage research, cooperation, and flexibility. Some suggested that the university should have both a European mission as an international center for the study of European cultures and an Alsatian one as a demonstration of the superiority of French culture to the region and to the world.[5]

The official reports stressed the need to outdo the Germans. The institutes and clinics should be continued; the library must receive both large government subsidies and donations from French publishers; and a new salary class above that of all other provincial universities and equal to those of the first class in Paris should be established. The Faculty of Letters was to put greater emphasis on modern languages and literatures than its German counterpart and less on ancient and oriental languages. Furthermore, in an attempt to make the university a model for reform, it was suggested that the Faculties of Medicine and Law be reorganized; it was proposed, for example, that the Law Faculty become the first Faculty of Social and Policy Sciences in France, following the German example.[6]

The new University of Strasbourg was inaugurated amidst great pomp and promise on November 22, 1919, just a year from the day on which French troops had first entered the city. Presided over by Raymond Poincaré, the ceremonies were attended by over 2,000 people. In attendance were the Faculty and officials of the new university, prominent Alsatian officials, and the military marshals, Joseph Joffre, Ferdinand Foch, and Henri Philippe Pétain. In addition, impressive delegations were sent from France's leading academic institutions and from those of the "allied" and neutral countries. In his keynote address Poincaré proclaimed that the university was to become a national school serving the homeland, a universal school open to all sciences and searching for their "fundamental unity," and a regional school taking into consideration the particular needs and aspirations of the region.[7]

The new University of Strasbourg was in fact exceptional in many ways. For a French provincial university, its sheer size was impressive. It had the highest number of faculties for any French university and the size of its Faculty of Letters exceeded those in all the other French provincial universities.[8] As advocated by official reports, modern languages and literatures (particularly French) were exceptionally well represented. The philosophy section included chairs for not only philosophy and the history of philosophy but also for psychology (for which the Germans had left impressive facilities) and for sociology, the first chair to be so named in all of France and one of only four stable sociology chairs during the interwar period.[9] A precedent had been set by the German university where Georg Simmel had taught courses on sociology, though officially a professor of philosophy. In any case, Pfister felt that given the progress made in sociology "under teachers like Durkheim," the field must be represented.[10] The number of positions in history and the associated fields of geography, archeology, and other "auxiliary" subjects was also exceptional. History's proportional representation within the Faculty of Letters (when counting the *maîtres de conférences* and the *chargés de cours*) was larger not only than Toulouse and Lyon but also than Paris, due mainly to the very strong representation of regional history. Also, as in Paris and Aix, though nowhere else, there was a post for religious history – a controversial one granted in part to counteract the presence of the theological faculties which were part of the university as a continuation from the German university.[11]

The Faculty was also distinguished by the academic inclinations of the new appointees. The historians were very open to social and economic history, as demonstrated by the presence of Bloch, Febvre, André Piganiol, Ernest Cavaignac, and to a certain degree Pfister. In addition the Faculty was well disposed toward the field of Durkheimian sociology. Maurice Halbwachs, a major spokesman for Durkheimian sociology, held the chair of sociology. During the inter-war period, he became an active critic of other disciplines, playing a role similar to that of Simiand in the years before the war.[12] Among others at Strasbourg who had shown an interest in the field were Bloch, Febvre, Piganiol, Pierre Roussel, and Charles Blondel, who, while at times critical of Durkheim, had followed his work closely.[13] Human geography, on the other hand, was not very well represented. Henri Baulig, who was in charge, was primarily a physical geographer and, though initially trained by Vidal, had later concentrated on geomorphology after studying with William Morris

Davis at Harvard. Maximilien Sorre, a young human geographer, helped Baulig to organize the geography program but returned to Bordeaux by December, 1919.[14]

Size and composition alone do not fully account for the Faculty's exceptional character. For those involved, one of the most remarkable features was the presence of the numerous "institutes" and "centers," many of which began in an attempt to take advantage of the facilities for "seminars" left by the Germans. As Pfister described: "Among all the Faculties of Letters in France, that at Strasbourg has the distinctive feature of being divided into Institutes. Each discipline has at its disposal a special work room, open from eight in the morning to seven at night and possessing all the current works which the students can take from the shelves themselves."[15]

The institutes did not meet, however, all the high expectations. To men such as Henri Berr and Charles Andler this use of the term "institute" was misleading. These "institutes" were simply "laboratories" and "workshops" not unlike those attempted at the Nouvelle Sorbonne rather than representing the ideal of "a grouping of Institutes where the specialists would call on each other according to the research of each and to the changing needs of science."[16] Though perhaps successful in promoting serious research, they were bound to fail in breaking down barriers between disciplines and even faculties, leading not to synthesis but instead to excessive specialization. How could it be otherwise, asked Berr, when there were seven "institutes" of history for eight professors?[17]

Much closer to the ideal of an institute promoted by Berr and others were the centres d'études, which were to serve not only professors from different disciplines but also from different faculties. Here at last, according to Berr, was a recognition of the need to bring together and foster exchanges between the disciplines. The first two of these centers organized, that for modern studies (covering the Renaissance and Reformation) and that for medieval studies, were in fact directed by two of Berr's collaborators, Lucien Febvre and Marc Bloch. A third center, for ancient studies, was added by 1921. The following description of one of the goals of the Centre d'Etudes Modernes, quoted by Berr, gives some indication of their spirit:

Give birth to and develop in the minds this fundamental idea . . . that one can not study one of the aspects of a civilization without knowing, at least cursorily, the others: that science is a collective work in every sense of the word – a cooperation and a coordination of distinct, but converging efforts. The center thus does not address some category of specialists (historians, literary persons, philologists, jurists) to the exclusion of others. It is intended for all of them at the same time.[18]

Strasbourg did in fact have an exceptionally cohesive faculty, a clear contrast to the situation in Paris.[19] With few local connections, without the distractions of Paris, and with comparatively light teaching loads, they turned to each other. Not only did they depend on each other for social occasions, they also exchanged ideas through the centres d'études, attended each others' courses, occasionally taught together, and even developed a system of inter-disciplinary seminars known as *les réunions de samedi.*[20] Beginning in January, 1920, with meetings for "philology, orientalism, and archeology" and for "history of religions," *les réunions de samedi* soon expanded to include meetings for "social history" and for "literary history." Professors from the various faculties and local lycées and occasionally visiting scholars would meet on Saturday afternoons in the appropriate institutes.[21] Claiming that this attempt at intellectual collaboration was both original and successful, the *Bulletin* of the Faculty of Letters noted, "They are composed of unrestricted discussions, during which the colleagues present inform each other on the most recent publications and discuss their value. Each thus keeps its own look and originality. They have nevertheless a family resemblance."[22] Lucien Febvre, when inviting the Belgian historian Henri Pirenne to a meeting of the social history group, explained that in addition to discussions about particular books, more general topics were also covered with an emphasis on methodology and "general problems."

for example, last year at one of our meetings, I put forward your ideas on the Evolution of Capitalism and the alternation between periods of freedom and of control; a discussion follows which is sometimes instructive enough because in attendance are our colleagues, the legal historians, whose formation and experience differs from ours – a few at Strasbourg are eminent, Perrot for example, or Chipeaux (sic).[23] There are also the philosophers, the literary people, and the sociologists; in brief, people of very different spirits.[24]

Although all these forums were attended by professors from several disciplines, their characters did vary. During the 1920s and early 1930s, the meetings for "linguistics, orientalism, and archeology"[25] and for "history of religion" were by far the most active. On the other hand, the "social history" seminars started by Febvre and Bloch petered out in the mid-1920s after a very successful beginning that had drawn on a wide range of participants from both inside and outside the Faculty of Letters.[26]

Some meetings were little more than accounts of recent publications and works in progress (including those of the participants), but others were characterized by methodological discussions and debates. The "linguistics, orientalism, and archeology" group, for example, demonstrated a keen interest in developments in linguistic theory, particularly that of Meillet, whose work was discussed in many meetings and for several years. In the short-lived forum, "social history," interesting exchanges took place over the methodology of history and geography. These included a debate involving Piganiol, Febvre, Baulig and Halbwachs over the relative merits of geography and social morphology, and disagreements over the economic theories of Karl Bücher between Pirenne and Halbwachs, and over the work of the German historian Alfons Dopsch between Bloch and Ernest Champeaux. Meetings within the literary history group, although somewhat less controversial, included discussions of the relative merits of "positivist theory" and "romantic theory" in interpreting popular poetry, of the possibility of a legal or even a social interpretation of the Nieblungen, and of the value of a comparative approach to literature.[27]

Like the meetings on "social history," those on the "history of religions" drew on quite a range of participants, who kept each other abreast of archeological findings as well as publications. It was there that many of the discussions of the more sociological works took place. For example, Blondel reviewed Lévy-Bruhl's *Ame primitive* and also spoke more generally on Lévy-Bruhl and Auguste Comte. Halbwachs promoted the works of Granet and Robert Hertz (both of whom had been influenced by Durkheim), and attacked Ernst Cassier for going against Durkheim's theory of religion and Charles de Rouvre's book on Comte for misrepresenting Comte's view of Catholicism.[28]

With time, the *réunions de samedi* became a bit more routine, characterized more by exchanges of information than of viewpoints and with fewer active participants. Nevertheless, even in the later years they served to keep their participants abreast of new developments in a number of fields and to familiarize them with work done not only in France but also abroad. The large number of discussions of foreign works and international conferences demonstrated that the participants were committed to encouraging international collaboration and communication among academics.[29]

In all, the atmosphere at Strasbourg in the early years was a challenging and dynamic one. In many ways, the reforms suggested and to a certain extent implemented were like those which Bloch promoted in his own writings on the Université. Modern languages were well represented, economic and religious history were both included and interdisciplinary and international collaboration were promoted.[30] As Henri Berr, who knew a number of the faculty there personally, described:

Happy are the professors of the University of Strasbourg! They have youth, for the most part; and the older ones have the renewal given by settling in a promised land. They have material resources which no other French university possesses – except for Paris . . . They have the feeling that their influence can reach forth into the distance, that the University of Strasbourg – like the Strasbourg cathedral – is destined to dominate large spiritual horizons, to "link up" spirits in a religion of the truth.[31]

More broadly, the careers of these men were filled with promise not just because of their location at Strasbourg but also because of the unhappy recent past. The First World War had taken its toll on both the older and younger generations of France's academic world. Both Vidal de la Blache and Durkheim were now dead (1918 and 1917 respectively) and a generation of historians were also reaching the end of their careers. Monod had died in 1912 and Lacombe in 1919. Ernest Denis would die in 1921, Lavisse in 1922, and Gustave Bloch had retired in 1919 and would die in 1923. Langlois and Seignobos still had some active years left but no longer commanded the following and attention they once had. The disciplines of geography, sociology, and history no longer had secure leadership.

The younger generation suffered more tragically from the war. Of 234 students in the ENS *promotions* of 1912 to 1914, 92 had died between 1914 and 1919, most of whom were killed in action. Even those who were somewhat older fared poorly; 10 out of 41 students in Bloch's *promotion* of 1904, for example, died between 1914 and 1919, men who would have been in their late twenties and early thirties.[32] Given the facilities and atmosphere at Strasbourg, the faculty there had an opportunity to make a real impact – not just within their own university but also in shaping the future character of their disciplines in all of France.

An outstanding example of the way in which the members of the Strasbourg faculties were able to take advantage of their new position was the founding of a new journal, the *Annales d'Histoire Economique et Sociale*. That journal would eventually become a powerful force in shaping work not only in history but also in many of the social sciences, as well as increasing the prestige and influence of its co-founders – Lucien Febvre and Marc Bloch. Many would be intrigued by its focus on economic and social themes at a time when reparations payments, fluctuating currencies, and the economic crisis of 1929 all demonstrated the power of economic forces.[33] Although publication did not begin until 1929, plans for the journal began soon after the reopening of the university. In many ways the *Annales* embodied the spirit of the new university during its early days of hope and promise – arguing for interdisciplinary work and for international collaboration.

According to Henri Baulig, Febvre began to plan for an "international review of political economy" as early as 1919, inspired in this project by Henri Pirenne. By April 1921 plans were well advanced and Febvre and Bloch approached Pirenne for the first time to gain his backing.[34] As Febvre explained to Pirenne, their new journal could help to overcome the fragmented and unsatisfactory character of work in economic and social history in France, which stemmed in part from the peculiar institutional structure of the French university system. One of the initial aims of the journal was in fact to establish closer links between history and sociology. As Febvre explained, there was a divorce in France between political economy, which was taught in the law faculties as a study of doctrines rather than facts, and history, which was taught in the faculties of letters, and was an antiquated discipline lacking a clear conception of its purpose, methods, and scope. What little economics was taught in the faculties of law tended to be dogmatic and to be used as an antidote against "bad doctrines." In addition Febvre explained:

In the Faculties of Letters, there is no training of economic historians. Sociology is in the domain of philosophers. Broadly, the historians know full well that one can not write a "history" without taking into account economic phenomena and social evolution. Or to be more precise, they only know this broadly: enterprises such as Lavisse's Histoire de France reveal this . . . [sic] cruelly enough sometimes (I say sometimes out of a concern for courtesy).[35]

As both Febvre and Bloch argued, their review was meant in part to fill an important gap caused by the recent war. Febvre reported that the *Revue d'Histoire Moderne et Contemporaine* had encouraged work in economic history and that when he approached them in 1914 to add a section on "the columns of social history," the editors seemed very willing. However, publication ceased when the war began. Although there were some reviews of political economy in France such as the *Revue d'Histoire Economique et Sociale*, they were run by law professors and accordingly interested only in the "history of theories" – not the sort of history that Febvre and Bloch hoped to promote. An additional impetus may well have been the cessation of the *Année Sociologique* due to the death of Durkheim and many of his collaborators. As soon became evident, Bloch and Febvre hoped to draw on some of the remaining Durkheimians, though Febvre wondered whether there were enough remaining Durkheimians to create a regular section on general sociology. Furthermore, the German *Vierteljahrsschrift für Sozial- und Wirtschaftsgeschichte* was no longer seen as a satisfactory outlet for historians in the "Allied" countries who had been among its contributors, and in any case its future seemed uncertain. A new international journal was needed.[36]

In their letters to Pirenne, both Febvre and Bloch asked him to become the director of the new journal. With the active collaboration of many in Strasbourg assured, they felt confident that strong support could also be found in Italy, England, and the United States, given the prestige and momentum that Pirenne's acceptance would lend the project. Pirenne replied positively to these pleas by promising his support, but remained reluctant to accept the directorship of the new journal.[37]

Time was of the essence since competitors began to emerge. In May 1922, Febvre wrote to Pirenne, expressing his worry about the reappearance of the *Vierteljahrsschrift für Sozial- und Wirtschaftsgeschichte*, and stressed again the need for a replacement in the "Allied" countries. He was also disturbed by the attempt of a recent Dutch journal *Tijdschrift voor Rechtsgeschiedenis* (founded in 1918) to become more international and observed: "It certainly seems that the idea of an international review is in the air."[38]

In 1923, the International Congress of Historical Sciences was held in Brussels; according to Bryce Lyon, Pirenne, as president of the congress, had ensured that German and Austrian scholars did not participate.[39] Febvre and Bloch presented their proposal for a new journal and explained that with the backing of the conference, they hoped to create a truly international journal.

However, their correspondence to Pirenne demonstrates that "international" in this case excluded the German and Austrian historians. This exclusion was sharply criticized by some American historians including, notably, Waldo Leland, an opposition that threatened support for the project from the Carnegie and Rockefeller foundations.[40] Their proposal and an alternative one by O. de Halecki, known for his Catholic interpretation of Polish history, were referred to a special commission so that Febvre and Bloch lost control of the project. Consisting of Pirenne, Febvre, Sir William Ashley from Britain, and Nicholaas Wilhelmus Posthumus from Holland, this commission met over several years.[41]

In 1927, Pirenne criticized the official report of the commission for emphasizing political history and down-playing comparative history. Funding for the commission's proposal was not found, and Bloch and Febvre went ahead with plans for their own journal without the backing of the commission. They were wary of the publishing house Alcan, who wanted to publish an "Année Economique" similar in format to the *Année Sociologique* instead of a *revue*.[42] With the help of Albert Demangeon, they managed to convince Armand Colin (the publisher for the *Annales de Géographie* of which Demangeon was now one of the directors) to fund their journal. As originally proposed, their new journal was to be one of "economic and social history."[43]

They chose the next international history meetings to be held in Oslo in 1928 as the place to announce the publication of the journal and relied on Pirenne to successfully kill any remaining hopes for a competing international journal sponsored by the commission. For his part, Bloch distributed a prospectus for the new journal, announced the plans for "a national review with an international spirit" at the session on economic history, and worked the hallways to rally support. Although noting that strong support seemed assured from abroad, he observed that from within France they could count on considerable opposition in part from the jurists as well as from other provincial faculties.[44]

The gap which their journal could now propose to fill was no longer quite so big. In Germany the *Vierteljahrsschrift für Sozial- und Wirtschaftsgeschichte* was appearing regularly and in Holland the *Tijdschrift voor Rechtsgeschiedenis* was now well established and supported by numerous international scholars. The *Economic History Review* edited by Richard Henry Tawney and E. Lipson had begun in 1927 in England with the active collaboration of Sir William Ashley, Posthumus, and Pirenne, and in the United States plans were being made for the *Journal of Economic and Business History* associated with Harvard's Graduate School of Business Administration and backed by the Business History Society based in Cambridge, Massachusetts.[45] The Comité International des Sciences Historiques, formally established after the Brussels meetings, started to publish its own *Bulletin*, which was another potential competitor. Within France a second series of the *Année Sociologique* began in

1925 and in 1926 the *Revue d'Histoire Moderne* appeared, published by the Société d'Histoire Moderne. According to its editors, this journal would fill an "existing gap." It was designed, they explained, as a replacement but not a direct continuation of the earlier *Revue d'Histoire Moderne et Contemporaine* as it now had a new format and aimed at an active "international collaboration."[46]

Although nominally creating a journal of economic history, Bloch and Febvre aimed at a very broad interdisciplinary perspective with a strong social orientation, very much like the academic atmosphere which pervaded their professional lives at Strasbourg. In 1928, Febvre explained to Max Leclerc, of the publishing house Armand Colin, that the journal could help direct studies in social and economic history. He continued: "In addition by this same journal, we want to establish a permanent liaison between the groups of workers who more often than not are unaware of each other and remain shut within the narrow domain of their speciality: those who are strictly speaking historians, economists, geographers, sociologists, or investigators mostly pre-occupied with the contemporary world."[47] In a letter to André Siegfried (a prominent professor at the Ecole Libre des Sciences Politiques) written in January 1928, Bloch explained that he hoped the journal could show scholars that special preparation was needed to treat economic subjects and to show economists in turn that "history exists." As he elaborated, they intended to cover: "A field of study very broadly conceived. *We insist on the word social.* This is my emphasis (study of the organization of society, of classes, etc.) along with the word, economic . . ."[48] Here Bloch echoed Simiand's arguments at the turn of the century not only in stressing a social approach to economic history but also in charging that economic historians too often had little specific training for their work. In a letter to the future editor, Febvre also echoed earlier concerns over directing the work of the *sociétés savantes*: "there is a need to provide a scientific education for numerous workers: lycée professors, provincial scholars presently left almost entirely on their own resources with no support, no ties."[49]

Reflecting their interest in economic and social history, the proposed editorial board included not only historians who were interested in such an approach (Pirenne, Henri Hauser, Piganiol and Georges Espinas) but also Halbwachs, representing sociology, and Charles Rist, as an economist who was sympathetic to social interpretations. That they hoped to bring in other fields was illustrated by the presence of Albert Demangeon as a representative for geography (in addition to his strong ties to the publisher) and André Siegfried, who combined interests in political studies and geography.[50]

Not all involved in this project were comfortable with the strong links which Febvre and especially Bloch proposed to establish with sociology. In particular, their editor, Max Leclerc, was quite upset, noting that some of its practitioners had been attracted to socialism and militant politics; and he questioned the appointment of Halbwachs to the editorial board.[51] Febvre, in turn, replied that their new review was not to publish articles on "general sociology" since that was the domain of the *Année Sociologique* but rather to help historians in their own work on economic history to which the sociologists brought both a useful perspective and a knowledge of different sorts of sources. As he explained:

Such are the reasons that we are keen to include a representative of this French school of Durkheim, which exercised such a incontestable influence (sometimes positive and sometimes negative) on all the men of our generation. He [Halbwachs] is not there to do the work of a party or doctrinal man, but as an informant, and to a certain extent as a critic.[52]

The choice of a title for the new journal was determined in part by a process of elimination, but it also illustrated the developing views over the journal's character. As early as 1921, Febvre had noted that not only was the obvious title of *Revue d'Histoire Economique et Sociale* already taken but also that such a title prolonged the rather arbitrary linking of "economic and social" in the label: "it is not just the economic which explains and gives rise to the social." As an alternative, he proposed "Revue d'Histoire et de Sociologie Economique" but then added, "But wouldn't it shock certain minds a little?"[53] Whether he was worried about the reactions of the sociologists or of potential editors was unclear. By 1928, Bloch proposed the title, "L'Evolution Economique, Revue d'Histoire Économique et Sociale" – a title rejected by Leclerc and Demangeon as being "a little heavy" and as inappropriate for a journal. Leclerc proposed first that of simply "Revue Economique" (which Febvre argued was too close to that of the journal of the law professors) and then proposed that of "Annales Economiques: Revue Critique d'Histoire Economique et Sociale" as a parallel to that of the *Annales de Géographie*, which Colin already published. Pirenne objected that such a title was too ambiguous and suggested that the title of "Annales d'Histoire Économique" would rightfully highlight the historical character of the review. The final selection of *Annales d'Histoire Économique et Sociale* thus served to draw an analogy with the *Annales de Géographie* and to highlight the historical character. Furthermore, by retaining the epithet "economic and social," they indicated that links would be drawn to sociology without however solving the issue as to the character of such connections.[54]

In carving out a niche for themselves, Bloch and Febvre had to consider not only the title and object of the review but also its potential audience. Given the number of journals already serving university professors, it was hoped that additional support would come from not only the provincial scholars and members of the *sociétés savantes* but also from the public mostly interested in "current questions" – in other words, businessmen and colonial officials. Although members of the law faculties were unlikely to be of much help in making the journal of contemporary as well as of historical interest, the "colonials," including such men as Henri Labouret (the director of the Institut International pour l'Etudes des Langues et des Civilisations Africaines-Centre), and the geographers were seen as very willing to help.[55]

Because of all their competitors, the journal could no longer draw on as broad a group of potential collaborators as had first been hoped. Whereas in 1921 and 1922, they had anticipated active help from Davy, Simiand, Granet, and Georges Bourgin, these men were now committed to the second series of the *Année Sociologique* and only Georges Bourgin became an active contributor during the first five years of their review. Hauser, who had initially been proposed as a "bulletin editor" for a section on "the organization of work," was now a collaborator of the *Revue d'Histoire Moderne* and a number of other journals.[56] As a result of their networks and the increasing demands for articles and reviews from prominent French university professors, the major source of support for their venture came from the two groups from whom they had initially expected so much: collaborators associated with the University of Strasbourg and foreign scholars.[57]

Changes had also occurred in the suggested format. In his 1921 proposal, Febvre had argued that a useful review of economic history should include "above all articles of method." Nevertheless, in their opening statement in the new journal in 1929, Bloch and Febvre proposed to attack the schisms between related fields with substantive articles having some methodological interest instead of publishing specifically methodological pieces. As initially proposed, considerable space was devoted to "practical and technical information" which in the end included discussions of particular economists, historians, and "men of action," reports on conferences, expositions, and research centers, and information on archival sources, libraries, journals and other sources. In addition, a section on "inquiries" gave a more in-depth look at specific kinds of sources and the techniques associated with them, including private archives (particularly commercial ones), the study of prices, the organization of banking, and maps of field systems. They did not, as initially proposed, include a lengthy "current bibliography as complete as possible" on the model of the "Bibliographie Annuelle" of the *Annales de Géographie*. Nevertheless, they did include numerous reviews of various lengths in which considerable attention was paid to methodological questions. Their "problèmes d'ensemble" were lengthy review articles often covering a body of work; "les questions de fait et de méthode" were full reviews often running to three or more pages; and "les courriers critiques" were brief reviews grouped together under various headings, headings which were to remain very flexible.[58]

The flexible presentation and coverage of these reviews contrasted with Febvre's 1921 proposal. There he proposed both specific "Sections of general information" which were to be assigned to particular bulletin editors with general headings for sociology, geography, demography, technology, and various types of economies and economic doctrines (see Table 4) and "sections of national information" which could vary between countries. In all the format had become both more flexible and somewhat less ambitious; it allowed methodological discussions, but oriented them around specific historical topics. Furthermore, in an effort to promote interdisciplinary exchange, collaborators were often asked to comment on works outside of their discipline – a contrast to the earlier proposals that Sion review works in human geography and Simiand with some help from other sociologists cover sociology. In the end, Bloch and Febvre tried to get the collaborators from geography and sociology to review works by historians, demographers and others, while historians, often Bloch and Febvre, reviewed works of geography and sociology.[59]

One remaining issue related to the way in which Bloch and Febvre formed a niche for the *Annales* needs to be mentioned, if not solved – that of the relationship between the *Annales d'Histoire Economique et Sociale* and the *Revue de Synthèse Historique*, the journal in which both Febvre and Bloch got their start. Both journals promoted interdisciplinary work with a strong historical character; the *Revue de Synthèse Historique* favored more general methodological articles while in the *Annales* such issues were usually examined in relationship to a specific historical case – an approach perhaps taken (despite the earlier plans) to avoid direct competition with the *Revue de Synthèse Historique*. Another obvious contrast was the specific interest which the *Annales* placed in economic history, an approach which was only one of several discussed in the *Revue de Synthèse Historique*. According to Berr, it was this that marked the *Annales'* particular contribution. That journal was founded, he claimed "to shed light on an aspect of the life of societies which has stayed in the shadows for too long and to which Marxism has called attention."[60]

In any case, Berr's interest in his journal had shifted. In 1924 he had founded the Centre International de Synthèse, a research institution not only for historical sciences but also for natural science, and in 1931 changed his journal from the *Revue de Synthèse Historique* to the *Revue de Synthèse* – including series for "general synthesis" focussed on the history of sciences and for "historical synthesis." The historical articles which Berr would continue to solicit were those which focussed on the "big problems of history." As he explained, "as far as history . . . we will do our best to turn down any article that would be suitable for any historical journal."[61] Bloch and especially Febvre continued their close relationships with Berr contributing not only to his journal but also to his series "L'Evolution de l'Humanité," and to the discussions at the Centre de Synthèse.

Despite all the competition, the journal was eventually successful. Although starting with modest aims of a circulation of only 500, the publisher pushed

Table 4 *"Sections D'Information Générale" and "Sections D'Information Nationale" proposed by Lucien Febvre in 1921*

Sections d'information générale
 1 Sociologie générale (Simiand?[a])
 2 Géographie humaine (Sion)
 3 Evolution des techniques (X)
 4 Statistique et démographie (X)
 5 Economie des peuples primitives (Davy)
 6 Economie des préhistoriques (Grenier?)
 7 Economie des civilisations anciennes de l'Asie Occidentale et de l'Egypt (X)
 8 Economie des peuples de l'antiquité classique (Francotte?) (Piganiol[a])
 9 Economie des civilisations de l'Asie-Extrême-Orientale et de l'Inde (Granet)
10 Economie des civilisations islamiques (Gaudefroy-Demombynes[a] – mispelled as Godefroy-Demonbins)
11 Economie des civilisation slaves
12 Histoire des doctrines économiques du Moyen Age (Carlyle?)
13 Histoire des doctrines économiques modernes (Scelle and Rist)

Sections d'information nationale – France[b]
 1 Histoire de l'industrie, commerce, finance, économie urbaine, Moyen Age (Espinas)
 2 Histoire rurale de la France des origines à 1789 (Bloch)
 3 Histoire de l'organisation sociale médiévale en France (Bloch and others)
 4 Débuts de capitalisme moderne en France (Febvre)
 5 Les modes d'organisation du travail - France avant la Révolution, régime corporatif, compagnons, etc. (Hauser)
 6 Commerce, circulation, transports - France moderne avant 1789 (Letanconnaux)
 7 Société française dans ses rapports avec la vie économique – XVI–XVIII (Febvre)
 8 La Révolution française: les problèmes économiques (Georges Bourgin or Charles Schmidt)
 9 Questions industrielles dans la France contemporaine (Hubert Bourgin?)
10 Questions agricoles, idem (Augé-Laribé?)
11 Commerce, circulation, transport depuis la Révolution à nos jours (Marcel Blanchard)
12 Finances et banques dans la France contemporaine depuis la Révolution (Allix?)
13 La société française dans ses rapports avec la vie économique depuis 1789 (X)

Notes:
[a] HP: letter of Lucien Febvre to Henri Pirenne, Mar. 1, 1922.
[b] By Mar. 1, 1922, the "Sections d'information nationale – France" were expanded to fifteen sections and some titles and potential contributors were changed.
Name in parentheses = "Rédacteur du Bulletin" suggested by Febvre. (X) = no name suggested. ? = Febvre unsure.
Sources: HP Lucien Febvre, "Note sur l'organisation d'une Revue d'histoire et de sociologie économique," Dec. 5, 1921.

them to broaden the circulation to at least 800; the *Annales de Géographie* after all had a circulation of 1100. However, by September 1929 they had reached only around 350 including, however, 80 French libraries and 37 foreign ones. Under pressure from the publisher, they did their best to augment "the current part" to gain the interest of the business world, an effort which meant soliciting more help from their geographical colleagues. As late as 1934, they still struggled to increase the number of subscriptions. Nevertheless, the journal appears to have had a fairly large readership, due in part to the high percentage of library subscriptions.[62]

Bloch and Febvre threw themselves into the task of not only editing the review and searching for new subscribers, but also of writing articles, numerous reviews, and notes on conferences, research institutes, and so forth. Their collaboration was initially an active, fulfilling, and challenging one, convinced as they were of the novelty of their view of history. In September, 1929 as they still struggled to gain support and to work out their editorial system and positions, Bloch wrote to Febvre:

And we should not be surprised not to have succeeded at once. We are naturally prisoners of our habits and our milieu. To make an improved "Revue Historique" would not have given either one of us much trouble. In what we have criticized there is, deep down, a sort of intellectual revolution; it is difficult to adapt others to it and even to adapt ourselves. And a journal, like ours, is inevitably a continuous creation.[63]

Indeed the *Annales* did differ significantly from the *Revue Historique*, which remained the major journal of the French historical establishment. It highlighted economic and social history and the new sources and methods needed to promote it, gave greater attention to modern and contemporary history, and published work of social scientists, businessmen, and administrators along with that of historians.[64]

In 1919 when reporting on the new Université de Strasbourg, both Gustave Lanson and André Hallays boasted confidently of its "renaissance," and in many ways their hopes for the new university were justified.[65] The university not only survived the switch from German to French rule, but it went farther than any other French university in meeting the aims of the educational reformers. Its staff soon gained a well-warranted reputation for academic excellence, innovation, and initiative. The "renaissance" however drew to a close soon after the founding of the *Annales* in 1929. Given its distinguished faculty, an increasingly uncomfortable environment in Alsace, and dwindling support from the national government, the university soon lost many of its more active faculty members to Paris and with them went much of the enthusiasm and sense of commitment to the university.

By the mid-1920s, the temporary honeymoon between the Alsatians and the French government came to an end. In 1924, Edouard Herriot, brought to power by the Cartel des Gauches, announced plans to introduce all the republican legislation into Alsace, sparking numerous protests from both Catholic and Protestant leaders who sought to keep the special religious legislation in effect in the region. Soon afterwards a new journal, *Die Zukunft*, and a new organization, the Elsass-Loringische Heimatbund, started to promote the German heritage of the region and to rally autonomist sentiment.

As autonomist forces grew, the university came under attack for failing to promote Alsatian studies more effectively and for its assimilationist policies. The university's fraternities responded by manifesting their earlier links with Germany. Faculty appointments were criticized for favoring candidates from the interior over Alsatians. Another sore point was the relatively low participation rate of the university in the student and faculty exchanges with Germany following the Locarno talks of 1925. Under this pressure, the faculty did try to become more directly involved in Alsatian affairs, but the thrust was still towards assimilation.[66]

In the early 1930s, the Depression and the rise of the Nazi party in Germany added further strains. Motivated by fears of unemployment, students shifted to fields such as medicine and pharmacy. More direct threats to the faculty came with the government budgets for 1933 to 1935 which lowered the special supplements paid to Strasbourg professors. Further challenges to Strasbourg's distinctive status came in fights with the Ministry of Public Instruction over specific faculty appointments as well as the ministry's reluctance to fill some of the vacancies. To those in Strasbourg it seemed clear that they could no longer control their own affairs. On the international front the rise of Nazism in Germany first served in Alsace to dampen the attraction to Germany, but later led to a polarization and politicization. By the mid-1930s anti-Semitism had risen significantly in the region and students and faculty alike were divided into political factions destroying the atmosphere of close collegial relationships and common purpose. In early 1936, Germany remilitarized the Rhineland and in Strasbourg many began to fear either war, or the return of Alsace to Germany in an attempt at appeasement.[67]

For Strasbourg's successful professors, the worsening climate brought about a diminished commitment to their Strasbourg mission and a sense that the time had come to find new posts in Paris. Although relatively few left in the 1920s, by the early 1930s a veritable exodus began. Febvre left Strasbourg for a chair at the Collège de France in 1933, making collaboration between Bloch and Febvre much more difficult. Both felt that for the *Annales* to prosper, Bloch must find a position in Paris as soon as possible.[68] In 1934, Febvre wrote to Bloch that the "Strasbourgeois" were indeed "invasive beings" and forming a new colony in Paris as they maintained their earlier friendships in the capital, and by 1935, he commented "how few survivors are left from the fine team of the beginning."[69] Because of Strasbourg's reputation for good research facilities and its new role as a stepping stone to Paris, strong candidates were found to replace the departing staff, but the earlier optimism and commitment had gone.[70] Sharing the general disillusionment, Bloch wrote to Febvre in 1935, "And if I had the leisure and the inclination, I would write a little essay, 'How in about 15 years one killed an intellectual center.' I mean the Faculty of Strasbourg."[71]

Bloch had been affected very directly by the changes. His closest colleagues were leaving and the rise of anti-Semitism was not something which he could ignore.[72] Unfortunately, the time was not a particularly good one to be competing in the academic marketplace, despite initial appearances of open career lines. During this period of budget restraint, the number of teaching posts in the Faculties of Letters did not keep pace with the rapid increases in students and academic diplomas. The reopening of the University of Strasbourg accounted for a large proportion of the new posts added in the inter-war period so that for those attempting to gain posts later, competition was intense. History in particular fared poorly in part due to increasing competition from the Vidalians. In addition, the historians suffered from a crisis of conscience as to the nature of their field following sharp criticism from the Durkheimians. Only with the new retirement law of 1936 did positions begin to open up again.[73]

When Febvre left for a chair at the Collège de France in 1933, Bloch began to look for a new position in Paris in earnest, entering the competitions for the chairs vacated in turn by Andler, Jullian, and Simiand at the Collège de France.[74] The competitions for these chairs were quite intense with some of the strongest candidates being some of his other Strasbourg colleagues, notably Albert Grenier, Halbwachs, and Blondel.[75] Bloch was not appointed to the Collège de France, but was eventually successful in replacing Hauser in the chair of economic history at the Sorbonne in 1936. Undoubtedly, a number of factors contributed to Bloch's failure to obtain a post at the Collège de France. Some have argued that anti-Semitism played a role, and Olivier Dumoulin has suggested that during the budget crisis, Bloch's approach may have been seen as too innovative.[76]

Although the Strasbourg years came to an end for Bloch and his colleagues, they had been extremely productive ones. Once transferred to Paris, they became bogged down with numerous commitments, academic and otherwise, interrupted soon thereafter by the war. For many, the major accomplishments in these later years would be the completion of projects begun at Strasbourg or in any case, the working out of ideas first developed there. More specifically, it was during his Strasbourg years that Bloch did the bulk of his research and writing, concentrating first on his interests in the relationships between sociology and history and later returning to work closely related to the field of geography.

7

Kings, serfs and the sociological method

At the University of Strasbourg, Bloch found himself in a dynamic atmosphere at an institution that was more open to sociology, the study of religious phenomena, and interdisciplinary exchange than any other university in France. With an excellent appointment and his doctorate soon out of the way, he finally had the freedom to pursue his sociological interests in earnest. In substantive writings on the thaumaturgic kings and the transformation of serfdom, he began to employ some Durkheimian concepts and to explore such topics as rites, collective beliefs, social classifications and the role of language. As his career advanced, Bloch also had the chance to address some of the methodological concerns raised by such studies more directly – reviewing other works on social class, writing on the comparative method, and criticizing the work of particular Durkheimians. Throughout, he addressed many of the same issues that figured in the pre-war debates between sociologists and historians, although the Durkheimian school, decimated by war, no longer had the same presence.

During February, 1919, while on a trip with Charles-Edmond Perrin to the Lower Vosges, Bloch spoke of his research plans for the future. Both were still mobilized and on temporary assignment to the University of Strasbourg. Bloch told Perrin that once finished with his study of the peasants, he would turn to one of the consecration rite of the French kings at Reims.[1] The result, published in 1924, was *Les Rois thaumaturges*, his first major post-war work. Although he did discuss the consecration rite in this work, his focus was actually on a healing ritual associated with royalty in both France and England during the Middle Ages. Royalty were believed to have the power to heal those afflicted with scrofula, a disease thought to have been a form of tuberculosis often accompanied by glandular swellings in the neck. During the rite, the king would touch the afflicted, leading, it was believed, to miraculous cures. According to Bloch, the apparent miracle could be explained by the character of the disease in which the swellings would disappear for no obvious reason.

For Bloch, this topic provided a way of exploring the interests in religious

110

phenomena and collective beliefs that he had developed at the Fondation Thiers. He found the topic intriguing because the healing rite was one of the best documented of the supposedly supernatural phenomena, and he was apparently encouraged to pursue it by his brother Louis, a medical doctor.[2] Because of its rather bizarre character, historians often overlooked this ritual in their study of "monarchical ideas," and Bloch proposed to fill that gap. Rather than simply studying the ritual as a peculiar anomaly, however, he proposed to associate it with the other collective beliefs which supported it so that one could begin to understand what "the mystic royalty" really meant. Through the study of this rite, Bloch felt one could demonstrate the great gap between the "mentalities" of the present and of medieval times.

Despite a number of parallels to the work of the sociologists, Bloch's approach in this work cannot accurately be labelled as Durkheimian. Bloch himself did not view his work as one of sociology. As he explained to Georges Davy in 1924, "My *Rois thaumaturges* is just published. By deference I have requested a review for the *Revue Philosophique* from Lévy-Bruhl (the book, without being one of philosophy, in the technical sense of the word, nor even of sociology, seems to me to be too close to collective psychology not to be mentioned in that journal)."[3] Bloch also failed to label his work as one of social history, writing instead, "In sum, what I wanted to give here is essentially a contribution to the political history of Europe, in the broad sense, the true sense of the term."[4]

What Bloch meant by "political history in the broad sense" was a political history which was not simply a traditional one focussing on events, but instead one also informed by sociology. As he explained, he saw the question addressed by his work as double. On the one hand, the royal miracle should be attached to the ensemble of ideas and beliefs of which it was one of the most characteristic manifestations. On the other, one should try to explain its more particular manifestations – exploring why the healing rite happened only at a specific time and only in France and England even though it stemmed from a movement of thought common to a whole area of Europe. Elsewhere in the text he referred to the particular causes as "occasional causes" or as those which were the most fortuitous; the general conditions were variously called "profound causes," the expression of "profound and obscure social forces," and "currents at the basis of the collective consciousness."[5]

There is a clear parallel between Bloch's approach here and Simiand's 1903 attack on the historians which had made the distinction between "profound causes" and "occasional causes."[6] However, in contrast to the sociologist Simiand, Bloch argued that the royal miracle could not be explained without a careful examination of "occasional causes" which led to the establishment of royal healing rites at a particular place and time. In other words, he sought to explain individual phenomena, something Simiand had argued was not possible, and in that attempt proposed to draw on both occasional and profound causes, linking phenomena of different orders and thus once again going against Durkheimian principles. Although Bloch did not engage in a clear case of Simiand's "political idol," he still paid greater attention to political phenomena than Simiand appeared to advocate. Neither was his treatment of individuals a clear example of Simiand's "individual idol," since Bloch treated the individuals as fortuitous and as effective only given the necessary underlying social conditions. However, in contrast to Simiand, and in agreement with the historians Mantoux and Seignobos, he did argue that individual wills could not be ignored. He wrote: "To give a clear idea of its birth [the rite of touching] at a precise date and in a specific milieu, one must call on phenomena of another order which one could call the most fortuitous, because they imply to a greater degree the interplay of individual wills."[7]

To a certain degree, Bloch's book was an example of Simiand's "chronological idol," given its heavy reliance on narrative (tracing the growth, decline, and end of the rite) and its failure to identify the normal type of either the healing rite or the supernatural kings. It was not a particularly good example of that idol, however, as he emphasized "duration" and evolution over origins, stressed the influence of the milieu, and made use of a comparative approach. As he explained, the rites were not simply survivals, but instead had a vitality closely related to the places and times in which they were practiced and could only be understood in those terms. Although Bloch did spend considerable time trying to identify the specific beginnings of the rites, he did not attempt to locate the origins of the more general belief in "miraculous royalty."[8]

The comparative approach which Bloch adopted was essentially that of a comparative history of the particular rite in the two countries where it was practiced (France and England), rather than a comparison within a given social type or one designed to uncover the transition from one social type to another – as the Durkheimians would have advocated.[9] According to Bloch, the very nature of his subject matter required a comparative approach. As he elaborated, "happy necessity, if it is true, as I believe, that the evolution of the civilizations of which we are the inheritors will only become almost clear to us on the day when we will know how to consider it outside of the overly narrow framework of national traditions."[10] Beyond this, Bloch was not very explicit as to what he felt his comparative approach could achieve. In practice, although he did not attempt to formulate any laws, he did note that similar conditions in the two countries led to the adoption of the rites. In both cases a new dynasty was being established which had need of prestige and backing. At a more particular level, Bloch tried to show that the rite was copied by the English from the French, rather than simply being a coincidence. However, in the related example of the magic power of the "medicinal rings," an interpretation by coincidence, drawing on popular beliefs, was used. Significantly, in neither case did he invoke a general explanation by social type.[11]

Bloch focussed on comparisons between France and England, but did not limit himself to them. He made frequent reference to the Western European context not only to determine why only the French and English kings practiced the rite (the necessary particular circumstances were missing elsewhere) but also to trace the more general belief in the sacred character of the kings and the various ways in which such a belief was manifested. Passing references were also made to the Byzantine Empire and even to Japan and China, but Bloch expressed reservations as to the usefulness of such far-flung comparisons. He felt that at best they could only provide very general approximations and at times they could be very misleading.[12]

To illustrate his reservations, Bloch explained why he found Sir James Frazer's work of limited usefulness. According to Bloch, Frazer was wrong in reasoning by analogy that since Polynesian leaders could both heal and inflict illness, the same must have been true in Europe. As Bloch argued, "The comparative method is extremely fruitful, but only on the condition of not leaving the general; it cannot serve to reconstruct the details."[13] Accordingly, comparative sociology could help one to reconstruct certain general "collective representations" but even then one must remember that "these grand ideas, common to all humanity, or just about, have evidently received different applications according to the places and the circumstances."[14] Furthermore to label the healing rites as "primitive" told one little of interest since, "Historical reality is less simple and richer than such formulae."[15] It was, after all, the very strangeness of that reality that had attracted Bloch to the topic in the first place. Since it was the particular applications linked to the rich "historical reality" which he found most intriguing, Bloch turned most often to the more immediate context. For example, rather than turning to the Polynesian islands, Bloch argued that more could be learned by studying the early Germanic rulers, whose influence was increased due to the records of early practices which were recorded in the Bible.[16]

Parallels can be drawn between the work of the Durkheimians and Bloch's focus on religious phenomena and his treatment of linguistics. However, even in these cases, Bloch consistently drew close links to the historical context.[17] Religious phenomena were studied in relationship to debates of the times, identifying the actors and milieux as precisely as possible. In addition, political and even economic interpretations were noted, limiting an investigation of the more general questions that a sociologist might ask. Bloch did not engage in a detailed analysis of language, but he did examine a key grammatical change in one of the phrases associated with the healing ritual – rather than simply limiting himself to a study of images along the lines of literary history.[18]

Although Bloch's approach was not strictly sociological, he had little hesitation in relying on some of their concepts, as demonstrated, for example, by his examination of the rites associated with royal power. Without using the label of formalism, as in some of his early writings, he was careful in these discussions to note the importance of gestures and of objects invested with great symbolic powers. For example, he wrote: "The sacred acts, objects or individuals were thus conceived, not simply as reservoirs of forces capable of practicing beyond the present life, but also as sources of energy susceptible of an immediate influence from this earth."[19] Again, when studying the formalism of thought embodied in the rites, Bloch stressed their historical character, noting links to particular historical figures and events.[20]

In *Les Rois thaumaturges,* one of the clearest indications of Bloch's attraction to Durkheimian thought was his frequent use of such Durkheimian expressions as "collective consciousness" and "collective representations," terms, however, which he used uncritically and without definition. Despite his attraction to the term "collective consciousness," Bloch indicated in a 1921 article that he had some reservations as to the way in which it was used by the sociologists; he found their use of the term "too metaphysical."[21] Presumably, he hoped to adopt the word for his own purposes, divesting it of some of these "metaphysical" qualities. For this reason, perhaps, he appeared to favor the term "collective representations." Although the distinction he made between these terms is not entirely clear, it appears from certain passages that the latter were somehow more specific and less profound or fundamental. The belief that the royal sign was placed on the right shoulder was, for example, "a collective representation." In other contexts, however, "collective representations" in the plural form and "collective consciousness" appear to be almost interchangeable.[22] Bloch's rather loose use of these terms becomes even more apparent when one notes the numerous other terms which he also used in similar contexts such as "collective opinion," "collective ideas," "intellectual and sentimental representations," "mental representations," "mentality," and "profound psychological elements."[23]

In contrast to Bloch's usage, for Durkheim "collective consciousness" was clearly different from "collective representations." The former was the system of enduring beliefs of a society which remained relatively fixed from one generation to the next and were the same throughout the society. As the "psychological type of society," it was not, however, equivalent to the "psychological life of society" which also included "representations" associated with specific functions such as those of government, science, industry, and the legal system. Accordingly, the collective consciousness in more "advanced" societies was increasingly diminished as the division of labor became more pronounced.[24] Bloch, on the other hand, never characterized the collective consciousness as a system or implied that it was time-bound and changing, and he gave no evidence of believing that the importance of the collective consciousness diminished as the society advanced. By divesting it of its "metaphysical" overtones, he made it more or less equivalent to collective representations. Revealingly, Bloch never attempted to examine collective consciousness directly, writing instead, for example, of beliefs which penetrated into "the inner most depths of the collective consciousness" and "influence exercised on the collective consciousness by obscure reminiscences."[25]

Bloch was closer to the Durkheimian usage for the term "collective representations," but even here there were differences. In common, Bloch and the Durkheimians used the term to refer to collective beliefs, legends, and so forth.[26] However, to the Durkheimians those beliefs were particularly interesting because they represented or "translated," "the way in which the group conceives itself in its relationships with the objects which affect it" or, in another characterization, "the actual state of the society."[27] Accordingly, Durkheim and Mauss examined parallels between "primitive" classifications for the social groups and those for the world in which they lived, and Mauss and Beuchat tried to demonstrate that the opposition in the seasonal morphologies of the Eskimo was reflected in the categorization of the people and of the other important elements of their environment. Bloch, by contrast, was more concerned with tracing the origins and decline of *particular* beliefs and in noting the importance of both "profound causes" and "occasional causes" which supported them.

In another contrast, Durkheim stressed the difference between "collective representations" and "individual representations" insisting that the former had a life of their own, combining with and repelling each other, and forming syntheses.[28] Bloch, on the other hand, wrote in his concluding chapter, "We have just followed the secular vicissitudes of the royal miracle . . . we have done our best to shed light on *the collective representations and the individual ambitions which, mixing together in a kind of psychological complex,* led the kings of France and England to claim the thaumaturgic power and the peoples to recognize it" [my emphasis].[29] This mixing of "collective representations" and individual ambitions was clearly against Durkheimian principles.

Whereas the Durkheimians were more interested in characterizations at a societal level, Bloch was more concerned with social differences including those between social groups within a given society. At times, the groups he described in *Les Rois thaumaturges* were based on a notion of social class but at others more on political interests. His most common division was that between the "common people" on the one hand and the "theoreticians" and "scholars" on the other. At the height of the healing ritual, all the classes were, according to Bloch, represented in the crowds of the afflicted hoping to be healed including "people with little," nobles, and the religious orders. At later stages, however, the skepticism developing among the "elite" was not shared by the common people. Other divisions described included those between the apologists for the monarchy and "partisans of papal supremacy," those between Whigs and Tories, Catholics and Protestants, and those within the Catholic church in France. Throughout, Bloch took pains to disclose the changing character of their beliefs and pronouncements as related to their milieu.[30] His approach here might be described as the study of the "psychology of diverse social milieux at a given epoch," terms he later used in a review of Johann Huizinga's *Herbst des Mittlealters*, which he criticized, by contrast, for presenting a "general psychology of a whole epoch."[31]

Bloch never claimed that his work was one of sociology, comparing it instead to one of collective psychology. Clues as to just what he may have meant by this term can be found in his acknowledgments, in which he singled out two of his Strasbourg colleagues. There he wrote: "In particular, my colleagues Lucien Febvre and Charles Blondel will recognize so much of themselves in certain of the following pages that the only way to thank them is to indicate to them these borrowings taken, in all friendship, from their own thought."[32] Febvre was engaged at the time in his study of Luther, which Bloch later characterized as the study of "the pressure of the social milieu on a powerful individual spirit."[33] Similarly in his favorable review of Blondel's *Introduction à la psychologie collective*, Bloch noted that collective psychology was "the study of the collective element in the individual."[34] In his attempt to characterize the field of collective psychology, Blondel had drawn more extensively on Lévy-Bruhl than on Durkheim, praising his attempt to link psychology and sociology and to stress the very different character of thinking in "primitive" societies.[35] Like Lévy-Bruhl, Bloch had a keen interest in the *differences* between "mentalities," rather than any evolutionary progression which they were thought to represent, and for this reason Lévy-Bruhl was intrigued with Bloch's topic. Quite possibly, Bloch saw his work as one of collective psychology rather than as sociology both because of his interest in the mixing of collective representations and individual ambitions and because of the parallels with the work of Lévy-Bruhl.[36] Significantly, however, Bloch used the term "collective psychology" rather than individual psychology, two fields which to him were very different.[37]

What then did the sociologists think of Bloch's work? The review in the *Année Sociologique* was written by another of Bloch's Strasbourg colleagues, Maurice Halbwachs. In general, Halbwachs pushed for a more analytical approach to a broader topic with more links with the social context, and he was far less concerned than Bloch with the particularities of place and time. Echoing Simiand's earlier attack on the chronological idol, Halbwachs suggested that rather than tracing the beliefs from their obscure origins through to their decline, one should concentrate on just a part of the period when they were fully developed. He argued that both Bloch's narrative approach and his narrow definition of the topic hindered the important links to the other customs and ways of thinking of the times. Rather than study the thaumaturgic kings, Bloch should have chosen either to study royalty or the miracle. In the first case, he could have done more with the parallels between "monarchical loyalty" and the "vassalic loyalty" giving a more thorough look at the collective psychology of the feudal period. If the miracle was the subject of interest, a more general study of the "theory of the miracle" would have been required; one would then include all sorts of miracle healings, be they related to kings, relics, tombs, or whatever, as well as a variety of other magical and religious practices.[38]

The French and Belgian historians who reviewed *Les Rois thaumaturges* emphasized the difference between Bloch's approach and that of James Frazer, and in contrast to Halbwachs praised Bloch for limiting his comparison to a particular historical context.[39] For example, Christian Pfister, Bloch's thesis supervisor, wrote in the *Journal des Savants*:

Monsieur Marc Bloch refrains from creating a work of sociology or comparative mythology. He does not want to walk in the footsteps of Sir James Frazer. He does not take the healing power of the kings back to the most ancient beliefs, to those of which we still find traces in the most backward populations of Oceania . . . It is thus a question of France and England in this volume, as the subtitle indicates very well: it is a comparative study of the institutions of the two countries, or the characters of the two peoples.[40]

The Belgian medievalist François Ganshof agreed that the work was, thankfully, one of history, not comparative sociology, and in the *Revue Historique*, Charles Guignebert, who held the chair of religious history at Paris, also noted and approved of the contrast to Frazer's approach.[41]

Other historians chose to highlight the innovative character of Bloch's approach. Henri Sée and Henri Pirenne both praised the interdisciplinary character of his work, and Guignebert and Gustave Lanson were also struck with the breadth of his explanations and particularly liked the connection which he made between the coronation rite and the early beliefs recorded in the Bible. Furthermore, Lanson, for one, noted the "psychological side" of Bloch's book. By contrast, Robert Fawtier argued that Bloch had attempted too much and would have been better advised to limit his study to one of the miracle of scrofula in the Middle Ages. Topics outside the realm of history should have been left aside, and for Fawtier that included both the development of the sacred character of the monarchy and "a sociological study of the beliefs of the popular masses in the Middle Ages."[42] Bloch in turn replied, "He [Fawtier] knows precisely where the Middle Ages and 'historical studies' end. I understand these limits less well and, in any case as far as history is concerned, I would undoubtedly not place the boundary marker – if I dared to place it – on the same line as him."[43]

Despite the praise for Bloch's historical approach and creativity, a number of historians had mixed feelings on the importance of Bloch's subject. After a few initial reservations, some such as Guignebert and Febvre felt that it was indeed very significant and worth the lengthy treatment which it received. Not unlike Halbwachs, others, such as Ganshof and Ernest Perrot (from Strasbourg's faculty of law), felt that it would have been more appropriate to broaden the focus to that of the sacred character of royalty. From the Ecole des Chartes came a somewhat mixed response. Maurice Prou wrote to Bloch that the book was interesting and to be admired for its use of very dispersed and unlikely sources. Gustave Dupont-Ferrier, on the other hand, was more effusive claiming that Bloch's book was "as scholarly as agreeable" and that it succeeded in raising the topic to one of "great importance."[44]

Although remaining a work of history (and of political history at that), Bloch's book clearly demonstrated an interest in the work of the sociologists. He examined social phenomena and yet linked them to the individual in the world of politics and stressed their particular historical character. In all, he strove to give what he would call an internal picture.[45] His topic was that of a miracle, studied within its social context, but placed within a narrative framework. Although a comparative approach was adopted, it was limited to comparisons within a particular historical and geographical context and failed to use a clearly defined typology of societal forms. Despite his extensive research and his openness to a variety of approaches, Bloch's book was not very well received. It was criticized by Halbwachs for being too circumscribed and too historical and by some historians for treating at great length a rather trivial and peculiar topic. It would be many years before he attempted another major work so close to sociology.

In 1921, Bloch wrote to Georges Davy that he saw his work on the thaumaturgic kings as simply an interesting diversion and that he needed to find the courage to return to his study of the rural populations of Ile-de-France.[46] Three years later in a letter to Berr, he still wrote of plans for that major work – plans however which were never carried out.[47] Instead he published several articles on the transformation of serfdom which drew heavily on the research completed for this unfinished work. Reflecting the shifts in thinking which he had made while still at the Fondation Thiers, Bloch chose to examine serfdom's long-term transformation as a social institution, rather than simply studying its disappearance, and also paid particular attention to religious ideas as agents of change. Thus, links to slavery, later developments of serfdom from a personal bond to an inferior social class, and the eventual entrance of some former serfs into the nobility were all explored. Although many of his examples were taken from the Ile-de-France, he never used that region to define his topic but focussed instead on more general developments of the institution both as experienced in northern France and as compared with southern France, England, Germany, Spain, and Italy.

Bloch published the six articles related to this theme during the 1920s, four of which looked specifically at the evolution of terminology and semantics. More than ever, he made use of a linguistic approach so that his method came closer to those of Meillet and at times, of Gernet. As Bloch explained, such a study was of interest since, "To clarify the history of words is to throw a brighter light on the things which they designate, or which they hide." In particular he hoped his work would contribute to the formulation of a future "lexicon of French institutions." Language was, he argued, "the interpreter of the collective consciousness," but as he also noted, institutional changes often preceded linguistic ones so that interpretation was not always straight-forward.[48]

In general, he appeared to prefer to use the language of the times because of its close links to social life. For example, he attacked scholars such as Montesquieu and Fustel de Coulanges for using the term *serf de la glèbe* which misleadingly implied that such people were attached to the soil; instead, he argued that the term *hommes de corps* used at the time gave a far more revealing picture of a bond which was between men rather than between a man and the land he occupied. Bloch was also intrigued by changes in terminology. In his work on the term *colliberti*, for example, he suggested that by studying its semantic evolution the historian could trace the fate of the Roman *colliberti* (freed by a common enfranchisement) who gradually fell into serfdom and became known in France as *culverts*.[49]

Bloch's other articles focussed more directly on institutional evolution. Linguistic evidence was used simply as one of many sources to address more general questions including the role of religious ideas. Although Bloch cautioned that the importance of "representations of a religious order" was difficult to weigh, he did note that one of the defining characteristics of a serf was his inability to enter religious orders. To their contemporaries, serfs lacked the independence necessary for servants of God. He also argued that religious factors were important in explaining the decline of slavery which allowed serfdom to develop. Christianity forbade its members to enslave fellow Christians, making recruitment more difficult, and also encouraged "manumission" at a time when economic and political conditions also favored such a development. Bloch even suggested that rather than simply relying on economic explanations of religious phenomena, one should also study the economic results of religious phenomena.[50]

His major article on liberty and servitude was subtitled "Contribution to a study of classes," and here he specifically examined the concept of social class, articulating views which had previously remained implicit in writings such as *Les Rois thaumaturges*. For example, he argued, "As human institutions are realities of a psychological order, a class only exists according to the idea one forms of it."[51] Bloch's approach, which stressed "the psychological order" and treated serfdom as "a collective notion," was not unlike those of Halbwachs and Simiand who stressed that class should be approached through the study of collective representations. Halbwachs, for example, had written in 1905, "what matters to the sociologist is the states of the social consciousness."[52] Similarly, the previous year, Bloch had endorsed the treatment of social class in Simiand's *Cours d'économie politique*:

"In a general way," he says in a phrase that historians will have to remember, "classes constitute themselves according to that which is most appreciated in a society, that which gives respect, that which entails power, authority, and the means of life and action, either real or considered as such." Let us underline these last words: "considered as such." They evoke a fundamental theme, which, I believe, is never actually expressed in the book – since what we have here is not a methodological work – but which is underlying throughout: the idea that social life is, above all, a matter of collective representations.[53]

Simiand, he claimed, was too good a sociologist to confuse the notion of class which focussed on "collective representations," with "material interest."[54] Given this interest in collective representations, it is not surprising that Bloch was so attracted to linguistic analysis.

In his substantive argument, Bloch concluded that serfdom was, in fact, a class institution since serfs as a group were treated by others as a separate class; for example, they were perceived as inappropriate marriage partners. Unlike the other groups tied to their superiors by links of dependence, the serfs were born into their position and as a result were seen to be deprived of liberty. Serfdom, however, could not be understood simply in class terms; it had a double character being both "a link from man to man" and a class institution. According to Bloch, the class character of serfdom gradually grew and the personal bonds weakened, so that by the thirteenth century a seigneur who elevated his serf to knighthood automatically freed him. Also, he noted that perceptions as to the specific restrictions which serfdom entailed, varied with the social milieux. The jurists, for example, showed much more certainty as to the principles involved than the "common man."[55]

Bloch's examination of the changing character of serfdom led him to several related themes, including that of *la ministérialité* explored in an article in 1928 and that of the rise of the nobility discussed in a 1936 article. *La ministérialité* referred to those who served their seigneurs in some special function, be it as a domestic servant, an officer of the court, or an overseer for the estate. Such posts could bring with them considerable wealth, power, and a very different *genre de vie* from that of the ordinary serf from whose ranks these men often came. As a result, *la ministérialité* eventually contributed to the formation of a noble class. According to Bloch, their *genre de vie* was more important than any "juridical condition" in that transition.[56]

In contrast to many historians (such as Sée) who relied on legal definitions, Bloch argued the "collectivity of the powerful" which preceded the appearance of the nobility constituted "a social class" – not a "legal class." As it stemmed more from opinion than law, that "class" was very difficult to pin down; it was poorly expressed by contemporaries, indecisive and flexible. Here, once again by stressing perception over law, Bloch adopted an approach more akin to the sociologists.[57] Further complicating matters, Bloch argued that this "the collectivity of the powerful" was comprised of several social groups. Accordingly, the investigator needed to proceed milieu by milieu, describing, "the patrimonies, the occupations, the armaments, the habitat, the alliances, the culture, the way of thinking, this *genre de vie* in a word, which more than any other characteristic, perhaps, constitutes the cement of a class in the process of formation."[58] In spite of the importance he attributed to class analogies, Bloch cautioned against a retrospective interpretation which tried to explain all of medieval society in static class terms. As he had written in an earlier review of Halbwachs, "I do not know if he has sufficiently realized the relatively recent character – if one dare apply this word to the 12th and 13th centuries – of the appearance of a noble class, strictly speaking."[59]

Characteristically, the questions which most interested Bloch in his writings on serfdom and related themes were those directed to a study of the changing relationships between conflicting and competing social groups.[60] In a 1937 review, Bloch charged that the Dutch scholar Petrus Cornelis Boeren had given an artificial and static view of the social structure of Flanders due to his reliance on legal terms rather than on the everyday language so crucial for an understanding of "the concrete" and "reality." According to Bloch, Boeren paid too little attention to time and misguidedly attempted to describe the "average serf," "a false average" of the type Simiand had attacked. As such his description lacked any meaningful relationship with "the beings of flesh and bone who are the real clients of history."[61] Bloch felt that the study of social class as well as other social classifications should be approached from the perspective of social history. Although drawing on both economic history and legal documents, such an approach should never lose sight of its aim, to understand social classifications as created and used by people of the times.[62]

Perhaps the most cited link between Bloch's approach and that of Durkheim is his advocacy and use of a comparative method. Closer examination of his methodological writings on the comparative method, however, can demonstrate both the significant difference between his approach and that of Durkheim and his changing views on this subject. These writings included a major 1928 article, which is still viewed as one of the key writings on comparative history,[63] as well as Bloch's contributions to the collective work on comparison for the Centre International de Synthèse, his reviews, and a statement of intent written when he sought a chair at the Collège de France.

Bloch's first major statement on the comparative method was presented at the 1928 conference at Oslo of the International Congress of Historical Studies. This was the same conference at which he "launched" the *Annales* and the two undertakings appear to have been linked. Both drew on Pirenne's work and pointed to the new form of history which Bloch and Febvre hoped to promote. His talk on the comparative method ended with a discussion of how comparative history could lead to a national history of an international spirit. He pleaded that the historians stop talking from the limited vantage points of national history, without understanding each other, but instead strive to base their terminology and methods on those used elsewhere. Bloch admitted that without local studies, comparative history was powerless, but also argued that without comparative history, local studies would lead nowhere.[64]

Despite Bloch's strong advocacy of the comparative method, his article was in fact very much the statement of an historian. As he confided to Henri Berr, many historians and sociologists held what he viewed as an "inconsistent" and undoubtedly out-of-date conception of history. Furthermore, the "big mistake" of the sociologists was to aim to constitute their science to the side of and above history, instead of refining the latter from the "inside." Unlike the sociologists, Bloch clearly hoped to promote an up-to-date version of history and to rebuild it from the inside – drawing on sociology in this process. This task was not without its institutional overtones, promoting history rather than sociology. On a more personal level, as Bloch would admit to Berr, he intended to use his article as an *exposé de candidature* for the Collège de France.[65]

In his article, Bloch developed a theme begun in *Les Rois thaumaturges*: the contrast between the wide-ranging comparisons associated with comparative ethnology and comparisons within a given historical context. Once again he showed a strong preference for the second but now was more emphatic as to the difference between the two approaches. The comparative method, he claimed, "is susceptible to two totally different applications, in their principles and their results."[66] In the first case, exemplified by the work of James Frazer, societies widely separated in space and time and lacking both mutual influences and a common origin were studied, a comparison which could only lead to very general observations. As he explained, "In brief, this far reaching comparative method is essentially a process of interpolating curves. Its postulate, and at the same time the conclusion to which it always leads, is the fundamental unity of the human spirit or, if one prefers, the monotony, the surprising poverty of intellectual resources at the disposal of humanity throughout history."[67] The second form compared neighboring and contemporary societies whose development depended on mutual influences, similar "major causes," and, to a certain extent, common origins. Bloch argued, "it certainly seems that of the two types of comparative method, that most limited in its horizon is also the richest scientifically. More capable of classifying with rigor and criticizing the parallels, it can hope to reach conclusions of fact that are both less hypothetical and much more precise."[68]

Bloch tried to articulate just what he felt the more "limited" and "scientific" approach could accomplish.[69] The most important accomplishment was that of highlighting contrasts between the neighboring societies and as such allowing one to determine their "originality." It could also help one to uncover important "historical phenomena" that were more apparent in one country than in another, and to study influences between neighboring societies. Yet another contribution was to help one discriminate between certain possible causes, determining their importance and avoiding the pitfalls associated with the "problem of 'origins'" – a problem which marred so many local monographs content to trace the developments without giving any real explanation. In exceptional cases one might uncover an unsuspected "community of civilization" between very different societies but, Bloch cautioned, such an attempt could be dangerous, as in the case of August Meitzen's debatable explanation of types of villages by ethnic backgrounds.[70] Finally, he suggested that through this approach one might find regions that were more appropriate to the study of "each aspect of European social life" rather than simply relying on those of political history.[71]

Bloch claimed to have borrowed "the general idea of development of the two forms of the method" from Meillet's monograph for the Oslo Institute for Comparative Research in Human Culture. An established linguist, Meillet was associated with the Durkheimians, having contributed to the *Année Sociologique*, but maintained a separate identity. In his monograph, he had made a similar distinction between comparisons which led to universal laws (demonstrating "the general unity of the human spirit") and those which gave one historical information (studying for the most part common origins but also at times mutual borrowings) – the type which Meillet preferred.[72]

Despite the obvious parallels, Bloch did not see the methodologies of historical linguistics and comparative history as equivalent. As he explained, "The history of social organization finds itself, in this respect, in a situation infinitely less favorable. This is because a language presents an armature which is much more unified and easier to define than that of any system of institutions. Thus the relative simplicity of the problem of linguistic filiations."[73] For example, Meillet noted that one had yet to find a case where the morphological system of a language resulted from a mix of the morphologies of two distinct languages; instead one found a continuing tradition with gradual changes. If, however, "real mixtures" were found, linguistics would, according to Meillet, need new methods. Bloch argued that it was just such mixtures which so often characterized the history of societies and, as a result, the search for a previously undetected "mother society" was destined to be almost always fruitless. Nevertheless, one could still hope to uncover interactions between human societies. Thus, in contrast to Meillet, Bloch stressed mutual influences over common origins, and while Meillet stressed "similarities," Bloch emphasized the detection and comprehension of differences. Although the forms of comparison used by these two men were similar, the aims were actually quite different.[74]

Bloch's conception of the comparative method contrasted even more markedly with that of Durkheim. In this article the only direct reference to Durkheim came in a footnote to Bloch's discussion of the ability of the second form of comparison to disclose false resemblances: he praised Durkheim's 1902 review of a work by Edouard Lambert, calling that review "one of the most accomplished methodological pieces that he has written."[75] Durkheim had argued that the "testaments" in ancient Rome and medieval societies were essentially different, the first serving the concentration of power in the family and the other "the dismemberment of the family." Durkheim concluded:

These two institutions, although designated by the same word, are thus, in reality, very different: they do not correspond to the same state of civilization, and in consequence, one cannot infer the history of one from the history of the other. These remarks apply to the comparisons which have often been made between the Roman family and the Germanic family. They emerge from two types.[76]

Bloch used this example to support his contention that the comparative method was particularly helpful in "the perception of differences." By contrast, Durkheim had argued, "In comparing the law of testament of the Romans and that of medieval societies, we fear that one brings together incomparable things."[77] To Durkheim this example appears to have represented a misuse of the comparative method, but to Bloch it was an example of its power – this despite the fact that Durkheim's example fell somewhere between his two types, being widely separated in time but not in space.

Several years later in his *Les Formes élémentaires de la vie religieuse* (1912), Durkheim came back to this issue when criticizing Frazer's *Totemism*. He argued both that "social facts" should not be separated from the social systems to which they belong and that comparisons could only be made between societies of the same "social species." Accordingly, social facts which appeared to resemble each other but which belonged to societies of differing "species" should not be compared. In his words, "The comparative method would be impossible if social types did not exist, and it can only be profitably applied within a given type."[78]

Durkheim's wording here indicates a rather different conception of the issue from that of Bloch, even though in his book he highlighted the comparisons between neighboring and contemporary societies (in Australia in this case) which Bloch later advocated. Theoretically, at least, comparisons within a "social species," as proposed by Durkheim, could range much more widely than those of Bloch's second category. Furthermore, their aims differed. When promoting an exploration of the mutual influences and common origins associated with a particular historical context, Bloch was more interested in establishing the originality of particular societies and in uncovering the richness of "historical reality," as he had argued in *Les Rois thaumaturges*. On the other hand, Durkheim, through his studies of similarities in societal organization, had hoped to shed light on the character of religion in general and even more ambitiously on the fundamental notions of thought common to all societies. Although it would not be accurate to label this approach as one which sought "the fundamental unity of the human spirit," the interest was more on the general character of social evolution than in an understanding of particular societies.[79]

Another way in which Bloch differed from Durkheim was in the role given to comparative analysis within his field of study. Bloch never went as far as Durkheim's early claim, "Comparative sociology is not a particular branch of sociology, it is sociology itself, as long as sociology stops being purely descriptive and aspires to account for facts."[80] By contrast, Bloch argued, "I do not come to you as a discoverer of a new panacea. The comparative method can accomplish a great deal; I consider its generalization and its improvement one of the most pressing necessities for historical studies today. However, it cannot do everything; in science there is no talisman."[81] For the historian, the comparative method should simply fulfill the role of a useful tool or technique, and as such it was well suited to "research of detail" despite its common association with the philosophy of history and general sociology.[82]

Bloch did not pretend to be the first historian to advocate a comparative approach but only to distinguish the two types of method and to demonstrate, through the use of examples, the value of the second. Nevertheless, he did have much on which to draw in the work of his fellow historians, and in his article acknowledged his debt to the writings of Pirenne, Berr, Sée, and Langlois. Pirenne had claimed that only a comparative approach could give a scientific understanding of "originalities." Berr attacked "crude use" of the method, and Sée promoted comparison across space over those in time. Langlois' characterization came closest as he advocated a comparison between medieval France and England based on their common origins and mutual influences, but in contrast to Bloch, Langlois had claimed that the main benefit was in uncovering parallels, rather than differences.[83]

In 1930, Bloch presented a report on the word "Comparison" for Berr's Vocabulaire Historique, a collective project undertaken at his Centre International de Synthèse. There Bloch kept his earlier typology and continued to show a preference for the second type of comparison but now claimed, in contrast to his Oslo paper, that both forms were equally legitimate and useful. Rather than charging that the wide-ranging comparisons implied "the fundamental unity of the human spirit," he wrote that this approach could demonstrate "the tendency of the human spirit to react, in analogous circumstances, in almost the same way."[84] As another example of his more conciliatory approach, Bloch even suggested, indirectly, that comparisons such as those done by Frazer had contributed to our understanding of the healing powers of the medieval kings, although "the phenomenon, as such" was quite different and not born until the tenth century.

Another modification came in Bloch's discussion of the benefits of the comparative approach. Whereas before he had argued that an attempt to uncover common origins was quite dangerous and only possible in exceptional cases, this time he pointed to work on the agrarian regime as an example where "unsuspected kinships" could be discovered even though one had to deal with "mixtures" rather than simple "filiations." Nevertheless, he continued to note that unlike language, other social phenomena "were not as well suited to research of this sort" and to suggest that the uncovering of differences and "originality" was the most important contribution which the method could make. However, he did add, in a more Durkheimian vein, that through an examination of such differences, "we can hope, one day to classify them [the social systems] and penetrate to the innermost depths of their nature."[85]

In the discussion that followed Bloch's presentation, he was criticized, particularly by Berr, for skimming over the importance of more general sorts of comparison – this despite the greater tolerance which he now showed for them. Berr also insisted that the uncovering of similarities was just as important as differences and that despite Bloch's intimations, there was nothing wrong with looking for "the identity of the human spirit." Bloch replied that he had stressed the differences primarily for practical reasons and agreed that there was a place for the study of similarities. Nevertheless, he remarked, "science uses the general in order to explain the particular." Berr, who now aimed at a synthesis of the various sciences, both physical and human, argued that the opposite was true, "the particular is only acquired in order to find the general."[86]

In a number of reviews during the 1930s Bloch expanded on the possible role for more long-range comparisons at which he had hinted in his 1930 report for Berr. For example, he suggested that interesting comparisons could be drawn between slavery in ancient Rome and Siam, that comparisons between Japanese and European feudalism could be very informative in uncovering both similarities and differences,[87] and that such comparisons might also shed some light on the question as to why capitalism appeared in Europe but not Asia. On the other hand, he saw little basis for a comparison of European and Chinese "feudalism." One must not, he argued, detach social phenomena from the milieu from which they were drawn. In addition, he insisted that one must be very careful in the use of concepts in such long-range comparisons particularly since historians often took categories such as "feudalism," "serfdom," and "tenure" from the European context and applied them "sometimes rather loosely to the entire world." For Bloch, any comparisons (and particularly those which ranged widely) depended on an effort to "tighten up the nomenclature" and on "the establishment of a rational vocabulary." In brief, though he eventually saw a place for long-range comparisons, he continued to insist, not unlike Durkheim, that there must be real grounds for such comparisons and that their success would depend on a very careful use of concepts.[88]

Bloch returned to this question in a statement of intent written when he unsuccessfully sought the chair left vacant by Camille Jullian at the Collège de France. To describe his field, Bloch chose the label "comparative history of European societies." It was a label to which he was particularly attached; he had used it in an earlier competition and was determined to keep it, even though some such as Simiand advised him to change.[89] In his statement, Bloch referred to his 1928 article saying that the principles expressed there "would form the basis of his teaching." Because of "common currents," he argued that a history of the European world could only be one of comparative history. As before, he stressed the ability of this method to uncover differences, but this time, rather than contrasting wide-ranging comparisons with those of neighboring and contemporary societies, he made a distinction between comparisons of parallel series (as he had attempted in *Les Rois thaumaturges*) and the study of "a restricted radius," if one "re-integrated" those observations into a larger ensemble. Implying that the comparison of parallel series could not often be successfully undertaken, he suggested that the method which he presently practiced, in *Les Caractères originaux de l'histoire rurale française* and in his study of serfdom, could have a more general application.[90] Although Bloch had softened his stance and came closer to accepting a Durkheimian use of the comparative method, the method which he practiced and the rationale which he gave were still those of an historian.[91]

To situate Bloch's relationship to the field of sociology more precisely, one can turn directly to his reviews of the post-war "Durkheimians." In comparison with Vidalian geography and history, Durkheimian sociology did not fare particularly well during the inter-war period. With claim to only four secure chairs and without even a shared *agrégation*, it still lacked an institutional base. Furthermore, the *certificat de morale et sociologie,* its main claim to legitimacy, represented only one quarter of the philosophy *licence,* and given the dearth of chairs, preparation for it was normally done under philosophers without strong Durkheimian links.

Sociology was an aging field with few new recruits as students increasingly turned to disciplines with a more secure footing, such as philosophy or history, at a time when fewer new academic posts were being created. Some continued to be interested in Durkheimian thought but, much like Marc Bloch, would practice within the framework of other disciplines and by so doing reinterpret and modify its methods and concepts. Even the *Année Sociologique* no longer acted as the intellectual center that it had been before the war. It did not resume publication until 1925 and then only briefly; only two volumes appeared, the second of which lacked the book reviews which had been so important before the war. The *Annales Sociologiques* which followed in 1934 was a very different enterprise. Divided now into five distinctive series, it illustrated the fragmentation and divisiveness which had come to characterize the field.[92]

As Johan Heilbron has so convincingly demonstrated, the inter-war "Durkheimians" were split into two groups which he terms *enseignants universitaires* and the *chercheurs.* The former were closely tied to the core of the Université in both teaching and administrative posts and were particularly interested in "general sociology." By contrast, the *chercheurs* or researchers pursued more "pragmatic" and empirical research and were much more leary of generalization and popularization. The first group, generally trained in philosophy or letters, was often associated with the republican establishment and the Parti Radical and tended to come from a somewhat higher social class. Its members included Bouglé, Fauconnet, Paul Lapie, Dominique Parodi, and René Hubert. The *chercheurs* were often socialist and had training and interest in fields such as statistics, economics, prehistory, and the history of religions. The institutional base of the *chercheurs* was not the faculties of letters, with the notable exception of Halbwachs, but instead the Ecole Pratique des Hautes Etudes, the Collège de France, and the Institut Français de Sociologie. This group included Mauss, Halbwachs, Simiand, Henri Hubert, Granet, Gernet, and Hertz. Given their empirical orientation, it was with them that Marc Bloch was more concerned.[93]

In his reviews, Bloch focussed his attention on the writings of two of the *chercheurs:* Halbwachs, who was closely associated with demography and statistics, and Simiand who did much of his writing and teaching in political economy. Although both were considered to be post-war spokesmen for Durkheimian sociology, even they had their public differences. For example, in a letter to Bloch, Febvre wrote of one of Berr's Semaines de Synthèse, "Long sociological session, talk by Halbw. full but very boring and more congealed than ever in a ludicrous Durkheimian orthodoxy. So much so that Simiand, intervening, had to clear the atmosphere (while apologizing for falling into heresy; these excuses being very amusing)."[94]

With only one exception (dating from 1925), all of Bloch's reviews of works by Halbwachs appeared in the *Annales*, and not surprisingly, given Halbwachs' active involvement in that journal, Bloch adopted a very positive tone towards his collaborator's work. He did, nevertheless, criticize that work for its tendency to be overly general and abstract. For example in his 1925 review of *Cadres sociaux de mémoire*, Bloch argued that Halbwachs gave a very vague impression of how collective memory actually worked, relying, despite Durkheimian principles, on arguments of teleology ("finalism") and anthropomorphism. Bloch suggested that the fault lay, in part, with the Durkheimian vocabulary which tacked the term "collective" onto works borrowed from individual psychology. Although endorsing the use of such expressions as "collective representations" and "collective memory," he proposed a more careful examination of the operations of collective memory to show how it differed from individual memory. For example, one could look at the transmission between the elderly and their grandchildren, errors in collective memory (as in Bloch's study of rumors), and changing legal custom.[95] Similarly, in 1929, when reviewing Halbwachs' study of Paris, Bloch argued that Halbwachs needed to penetrate more deeply into the social process, analyzing the subgroups and conflicts within "the Parisian collectivity," rather than treating it as a monolithic whole. He also criticized Halbwachs for using abstract terms like "pressure" and "tendencies" which were "stained with an involuntary teleology ('finalism')."[96]

By 1931, Bloch noted what he saw as a hopeful sign. In *Les Causes de suicide*, Halbwachs had used the concept of *genre de vie*, borrowing from geography. By so doing, he encouraged a more internal and less abstract view of human groups, not unlike Simiand's efforts at a new definition of social class and Mauss' work on civilization. As Bloch argued, one could not assume that "old multi-purpose terms" such as "nobility" and "bourgeoisie" could effectively encompass "all the human realities."[97] Similarly, in a favorable 1935 review, Bloch suggested that Halbwachs' *L'Evolution des besoins dans les classes ouvrières* marked a considerable advance over his earlier work as it went beyond static definitions to examine the diversified evolution. Bloch was pleased to report that this increasing attention to time was also apparent in the writings of others of "the same school."[98]

In his final review of Halbwachs' *Morphologie sociale* published in 1939, Bloch clarified what he meant by an internal study of human reality. He praised the work for always striving to look behind the material manifestation for "the social element, par excellence. I mean the mental element," and returned again to his argument that one should examine "realities" not abstractions. As he concluded, "There surely is the true realism . . . Because the "concrete," for the human sciences, is it not, by definition, the human."[99] To Bloch, human reality, while "concrete," was not simply material but more accurately the mental representation of material forces.

Bloch also commented in these reviews on Halbwachs' exclusionist conception of the field of sociology, which he saw as hindering fruitful interdisciplinary exchange. For example, in 1929, he criticized the opposition which Halbwachs attempted to establish between the "scientific" method of sociology and the historical method and remarked that this part of the Durkheimian legacy was perhaps not that most worthy of life.[100] By contrast, in his 1931 review in which he returned to this theme, Bloch criticized only Durkheim, referring to "the barriers which, we believe, Durkheim himself was wrong to want to raise."[101] This time, Halbwachs was praised for his interdisciplinary approach. Describing Halbwachs' work on suicide, Bloch remarked, "And by a curious contamination of the two great disciplines which one can say renewed, or ought to renew history, he added, borrowing a term from the vocabulary of Vidal de la Blache, 'that is what we call a *genre de vie.*' "[102]

Bloch never knew Simiand very well but had a high regard for his work. Accordingly, he was very keen to have that work reviewed in the *Annales*. In 1933, he wrote to Febvre suggesting that they switch places, with Bloch writing the review for the *Annales* and Febvre for the *Revue Historique*. As Bloch explained, "in order not to delay signaling to our readers what is certainly a considerable oeuvre (with however very debatable biases). Most historians will not read it. Nevertheless, it is advisable to give them an idea of it."[103] Despite differences in their approaches, Bloch continued to think highly of Simiand's work. In 1935, for example, he wrote to Febvre, "Just as I close this letter, *Le Temps* informs me of the death of Simiand. I knew he was ill (heart) from Halbwachs whose reticent remarks made me understand that it was serious. I was, you know, fond of him. In a word he was someone who accomplished something."[104] On the whole, Bloch was less quick to accuse Simiand of being overly abstract than he had been for Halbwachs, but even Simiand was criticized for not always taking a full account of human realities. Also, without directly charging Simiand with hindering interdisciplinary communication, Bloch remarked in a 1934 review, "the sociologist, the historian – I am one of those who sees no gulf between the two."[105]

In Bloch's first review of Simiand (1931), he praised Simiand's course on political economy, taught at the Conservatoire des Arts et Métiers. There, Bloch explained, Simiand successfully combined rational categories with a sense of the "concrete" in his portrayal of class and economic life.[106] By contrast, in 1934, Bloch charged that Simiand's *La Salaire, l'évolution sociale et la monnaie* was too detached from life, in part due to its heavy reliance on abstract labels as in "precept J1," "category A1," etc. More fundamentally, this abstract character stemmed from Simiand's refusal to consult what Bloch termed "documents of intention," "documents capable of informing us on the preoccupations of the human agent." As Bloch also put it, in order to understand his numbers, Simiand should examine, "the imprints of a different kind left by the desires, the fears, the prejudices of men, all these ideas or these sentiments in which we recognize unanimously the profound motors of history."[107] According to Bloch, if Simiand had examined the appropriate documents of intention, he might have been able to avoid his awkward position of having to rely on particular events, such as the discovery of sources of gold, to explain his economic cycles. Such documents, Bloch claimed, might well demonstrate that social conditions had motivated the explorations and as a result led to the discoveries. More generally, Simiand's reluctance to study these documents put the sociological approach in jeopardy. Bloch explained, "In truth, to consider only numbers as objective would not only force me to excommunicate in a block a good many of the aspects of sociology or history: notably religious sociology. It would also be fatal to economic sociology."[108]

Two years later, shortly after Simiand's death, Bloch praised his study of the United States for using such documents of intention as speeches, letters, and statements of motives. In addition, Bloch noted with approval that Simiand had paid greater attention to "differences" than before: "this preoccupation with the different, which is the very law of all history." More broadly, Bloch suggested that Simiand's work gave one hope that an "enlarged orientation" and "common agreement" were finally leading to "a science of men in society."[109] To believe Bloch here, Simiand had finally seen the error of his ways and realized that good history and good sociology were one and the same.

Behind Bloch's disagreements with Simiand lay Simiand's fears of teleology and his insistence that one avoid explanation by motives and goals. Bloch, on the other hand, had come to side with Mantoux and Seignobos in the pre-war debates. He argued that human phenomena could not be understood without a careful study of the conscious and unconscious motives associated with them – an issue he had skirted in his 1906 notes. Although Simiand eventually seemed willing to examine what Bloch called "documents of intention," he probably would not have seen them as expressions of motives, treating them instead as evidence of less personalized social forces.

Despite Bloch's un-Durkheimian stance, insisting on a direct examination of human motives, he found himself criticizing Simiand as well as some of the other post-war Durkheimians for belittling the touchstone of Durkheim's method, the comparative method. For example, in 1934 he challenged Simiand's assertion that because of the need for an "identity of base," it made more sense simply to study one society over time, something which did not accord with Bloch's understanding of the comparative method. Bloch argued that the comparative method was needed even to establish Simiand's "identity of base" and that furthermore, Simiand gave too little credit to the power of the method "as a reagent capable of setting aside the secondary conditions."[110]

In addition to his reviews of specific post-war Durkheimians, Bloch reviewed the attempts to revive the pre-war *Année Sociologique*. Despite their initial thoughts of filling the gap left by the end of the first series, Bloch was very supportive in these reviews of the efforts to revive the journal. For the new series, Bloch commended what he saw as a less rigid approach. He did, however, criticize a neglect of the "intermediate civilisations" (those between the primitive and the modern ones) so crucial for an understanding of present European societies. Bloch approved of the reorganization attempted by the *Annales Sociologiques* into a number of series, each devoted to a particular branch of sociology, but he found the four series which he reviewed of unequal usefulness to the historian and had the most praise for those dominated by Heilbron's *chercheurs*. Series D on economic sociology, which included contributions from Simiand, Georges Bourgin, and Halbwachs seemed particularly useful, and Series E, directed by Halbwachs, also had much to offer even though it tried to encompass the very disparate subjects of social morphology, language, technology, and aesthetics. On the other hand, Series C, "legal and moral sociology" directed by Jean Ray, had a tendency to be too abstract, and of Series A, "general sociology" directed by Bouglé, he wrote, "general sociology is connected to our studies by weaker ties than any other [branch of sociology]."[111]

Bloch approached sociological concerns at an early stage in his career, in his substantive work on the thaumaturgic kings and on the transformation of serfdom. Nevertheless, he was careful not to identify himself as a sociologist and soon became one of their critics, particularly following the founding of the *Annales*. In *Les Rois thaumaturges*, Bloch's interest in the relationships between the individual and the social became apparent, but he did not articulate just what those relationships might involve. Although he also demonstrated his attraction to certain Durkheimian terms, he failed to explain just how his usage differed from that of the Durkheimians. Though a study of a miracle, the work was not well received by Halbwachs, who criticized its chronological approach and its narrow topic.

The articles on the transformation of serfdom were somewhat closer to accepted sociological concerns. They examined changes of key social institutions over long periods of time, approached social classifications through the ideas people had of them, explored the changing meaning and forms of particular terms using techniques similar to Gernet, and stressed religious factors over political events where appropriate. However, in contrast to some of the Durkheimians, Bloch emphasized the great complexity of these phenomena and was very cautious in his generalizations. This was evident in his treatment of social class; he favored an examination of the *genre de vie* associated with particular groups and stressed both the dynamic character of social classifications and the varied relationships between many social groups.

Although drawing to a certain extent on both Meillet and Durkheim, Bloch's characterization of the comparative method in 1928 was very much that of an historian. Closely related to discussions among historians at the time, his depiction contrasted with theirs both in the detail of his analysis and in his very staunch promotion of the comparison of neighboring and contemporary societies to promote the uncovering of differences. With time he softened his criticisms of the long-range comparisons. Perhaps he was aware of the dangers of alienating others at the Centre International de Synthèse and at the Collège de France and was now familiar with some of the work that had been done, paying due attention to establishing proper grounds for more long-range comparisons. Gradually, as he accepted the possibility of such an approach and reacted against what he considered to be sloppy comparative history, he adopted a position somewhat closer to that advocated by Durkheim. He indicated the possibility of classifying societies; asked general questions on the character of slavery, feudalism, and capitalism; stressed the importance of relating social phenomena to their milieux; and dwelt on the importance of a careful definition of concepts. Nevertheless, he consistently argued that uncovering of differences was the main contribution which the comparative method could make to history and, increasingly, he dwelt on the theme of "originality."

His perspective, often implicit in his substantive writings and even in the articles on the comparative method, became more explicit in his reviews. There he attacked some sociological works for being overly abstract and pushed for a more "internal" and historical view of the problems which they addressed, insisting on a careful examination of "documents of intention." He was clearly interested in the role of human agency and unwilling to accept Simiand's early argument that one must avoid explanation of social phenomena by motives and goals.

Bloch's writing had come a long way from his early notes on the 1903 debate and from *Rois et serfs*. His chosen problems and approaches were now closer to those of the Durkheimians. At the same time, he had begun to clarify how he differed from them and to address, albeit often indirectly, the issues of teleology and motives. He remained very much the historian, fascinated with the richness of historical reality. Nevertheless, his version of history had come to fall clearly within the domain of social rather than political history.

8

Reflections on the geographical approach and on the agrarian regime

In his post-war writings on human geography, Bloch's position was first and foremost that of a critic rather than a practitioner – a contrast to his on-going attempts to write on sociological topics. Despite his many significant criticisms, he would argue in numerous reviews of geographical works that a geographical perspective had a valuable contribution to make to the type of interdisciplinary work that he found most satisfying. Eventually, through his efforts to write a synthesis of the agrarian history of France, Bloch became more directly involved in the use of geographical materials and methods and was himself a major subject of discussion and criticism by the inter-war geographers.

Despite Vidal's death in 1917, the inter-war French geography with which Bloch was engaged continued to follow the lead established by Vidal at the turn of the century. In part, this continuity was due to the power wielded by the first generation of Vidalians on academic juries. Characteristics in common included the preference for regional monographs, the important place given to physical features – especially those of geomorphology (a tendency strengthened by de Martonne's new leadership role), the stress on rural as opposed to urban or economic geography, a focus on France, and a reluctance to enter into serious methodological discussion. Although the approach remained much the same, the regions studied often became smaller, in part due to the increasing complexity of the disciplines on which they drew. However, some monographs, with a more thematic approach, still tackled larger regions. Also, given the extensive treatment of northern France before the First World War, southern France now received most attention.[1]

Published shortly after the war as part of Berr's series "L'Evolution de l'Humanité," *La Terre et l'évolution humaine* by Lucien Febvre directly addressed some of the pre-war controversies on the natures and domains of geography, sociology, and history. Given its polemical character, it was widely read and reviewed.[2] For the first time, Bloch took an active part in this debate, since he was in charge of the review of Febvre's book for the *Revue Historique*. There he noted that in tackling the subject of "milieu" Febvre was forced to grapple with the differing approaches of Vidal and Durkheim, about whom Bloch observed, "they have scarcely one trait in common, that is to have both marked a whole generation of historians with an imprint that will not wear away."[3]

Taking a stand somewhere between those of the geographers and sociologists, Bloch argued for an approach drawing heavily on sociology without, however, endorsing the Durkheimian view of science. Paralleling Halbwachs' review which stressed "representations" over "material agents," Bloch regretted the absence of a chapter on "the geographical illusion," which would explore how "representations" of the environment, such as the concept of natural boundaries and cartographic depictions, affected human conceptions and social life. More fundamentally, he argued that Febvre was misguided in giving unqualified support to the regional method of the Vidalians. For his part, Bloch suggested, "If, following the advice of Monsieur Simiand, a geographer had studied the distribution of dwellings in different well chosen *pays*, would he not . . . have released us from the quasi-superstitious importance attributed to certain causalities more quickly than many regional studies have done?"[4] Referring to his earlier discussion in "L'Ile-de-France," Bloch argued, "if in truth one must admit that the region is something so variable and so little given, does it not justify the fundamental criticisms of the sociologists to a certain degree?"[5]

On the other hand, Bloch approved of Febvre's discussions of possibilism and his hesitancy to talk about geographical laws, which served in the book both as mild criticisms of some geographical work and as a defence for its separate status. His review contrasted here with that of Halbwachs who charged that Febvre's position led neither to "a single really general law" nor to "any verifiable prediction."[6] Not being one to favor sharp disciplinary boundaries, Bloch failed to address the issue of the respective domains of sociology, geography, and history, which had been raised by the early Durkheimians and was also discussed by both Febvre and Halbwachs. Like Febvre, Bloch appeared willing to grant geography a modest role; it was, however, an even more modest one then Febvre envisioned, given Bloch's endorsement of Simiand's criticisms.[7]

Bloch received some support for his positions from Henri Sée, an historian who had done much to establish the field of social and economic history. As he wrote to Bloch:

Your observations on the book by L. Febvre appeared to be very sound. Basically history (especially economic and social history) must be even closer to sociology than geography; and the sociological method, as defined by Durkheim, is for the most part an historical method. We need monographs, and lots of them, but synthetic and comparative studies are no less important for the progress of science.[8]

Bloch's review of Febvre's book contrasted markedly, however, with those of Henri Hauser, Albert Demangeon, and Camille Vallaux, who, not unexpectedly, came to geography's defense – the former being an economic historian who had promoted the use of Vidalian methods by historians and the others, students of Vidal.[9]

Another project planned before the war but completed afterwards was the *Géographie Universelle*. Initially designed by Vidal and continued on his death by Gallois, the series did not begin publication until 1927.[10] In general, Bloch's reviews, which appeared in the *Bulletin de la Faculté des Lettres de Strasbourg*, were more critical than those of Febvre, which appeared in the *Revue de Synthèse Historique*. Although Bloch thought the works contained much of interest, he found that a number of these geographers still fell into the trap of geographical determinism and overlooked important social and historical factors, this despite the earlier debates between the Vidalians and Durkheimians. He also found that the geographical methods were not always used effectively. Bloch criticized the use of boundaries of the major states as the divisions in the series (an irony given the Vidalian interest in natural regions) and wanted more attention to synthesis and a better use of maps.

Bloch's most far-reaching criticism came in a review of Albert Demangeon's *Iles Britanniques* (1927), a work Febvre described as, "a book without weaknesses." By contrast, Bloch argued that the work paid too little attention to those historical factors that could help to explain Britain's agrarian revolution. More fundamentally, he suggested, "But how can one explain all of that in a geographical book? Perhaps the only course would have been to admit in plain language that at that stage any search for causality eludes the geographer; obviously on this point we reach the limits of the method."[11] It appears that in Bloch's view the geographical approach was at best a limited and primarily suggestive one, at least for the questions which he found most interesting. His highest praise was for Jules Sion's *Asie de Moussons* (1928). Bloch liked the balance struck between geographical factors and human ones, stressed the social character of the latter, and also liked Sion's ability to raise questions without jumping to conclusions.[12]

One of Bloch's concerns during the post-war period was to promote a reworking of the field of historical geography.[13] Despite his earlier praises for Longnon, he argued that historical geography as traditionally practiced in France had little to offer, writing of "this historical geography for which the collections to date, comprised of merely boundary lines, give us a simple carcass not living flesh, and, even at that, very imperfectly."[14] Nevertheless, he viewed the field as one of great potential and one which the *Annales* was well placed to serve. As he explained to Febvre in September, 1928: "Finally, have you thought about the means of utilizing our journal to benefit this historical 'geography,' which interests you with such just cause, and which lacks a mouthpiece (geographers despising it, and historians not understanding it)."[15]

Unfortunately, however, Bloch's published views on this subject were rather limited, in part because he felt that Febvre was better placed to make the sort of criticisms that were needed.[16]

According to Bloch, the field of historical geography must be defined by its object rather than simply as the use of early documents for geographical ends.[17] However, given his promotion of interdisciplinary research, he was never very explicit as to just what that object should be. He wrote, for example:

Perhaps one should still ask himself to what extent an "historical geography" of this kind differs from a history attentive to geographical factors, among others, and anxious to bring to its aid this admirable instrument of interpretation, which is the map – in other words, to history as it ought to be written. But for my part, I confess that I do not have a very strong interest in these problems of boundaries.[18]

Some of his most explicit remarks as to the object of historical geography came in reviews of the English geographers, Gordon East and H. Clifford Darby, who had come closer than the French historical geographers to Bloch's ideal. For example, Bloch praised *An Historical Geography of Europe* by East for its treatment of settlement and of economic and political geography. Bloch also claimed that *An Historical Geography of England before A.D. 1800* edited by Darby lacked a conceptual definition, but that its object appeared to be the relationship between the physical "substratum" and "human phenomena" with a stress on the "habitat," *peuplement*, forms of agriculture, and the location of industry and centers of exchange. Although endorsing these topics, Bloch suggested that the work could have been improved by an examination of the geographies of political opinion, schooling, and religion, i.e. it should have been broadened to include phenomena other than those typically treated by geographers simply because they were "the most immediately perceptible."[19]

In spite of Bloch's disdain for much of French historical geography, he praised the work of the French political geographer, Jacques Ancel, as a model for future work in the field. In his work on Central Europe, Ancel, according to Bloch, tackled the phenomena of "profound structure," including "the occupation of the soil," property relationships, the formation of social classes, and political power. Furthermore, he rightfully depicted the "state," "nation," and "frontiers" as being in perpetual movement. Although Bloch never expressed it in these terms, it seems that his vision of historical geography could be described as an historical study of social morphology, examining the relationships between "the substratum" and such social phenomena as the formation of social classes, religion, and schooling.[20]

When the *Annales* first appeared in 1929, researchers had just begun to show an interest in the "habitat." In his reviews of their work, Bloch pushed once again for greater attention to social and historical factors. In addition, he argued for comparative work and for more serious attempts at interpretation.[21] He suggested, for example, that Demangeon's classification of settlement types paid insufficient attention to important differences and changes in the social structure, often in turn related to changes in family structure.[22] Similarly in a review of Omer Tulippe's habitat studies he argued, "To count individuals and their residences is good; to count families would be no less useful; to describe the functioning of groups who live and toil together better yet."[23] He also criticized works such as those of August Meitzen which relied on ethnic interpretations of settlement forms – a result, he suggested, of research designs inappropriately limited in their ethnic domains.[24]

Of all the works by geographers on the "habitat," Bloch particularly liked those of the German geographer, Robert Gradmann. In 1930, Bloch praised Gradmann for his careful depiction of the historical stages of occupation and for his use of an interdisciplinary approach which helped him to avoid an ethnic interpretation for the distribution of dwellings. However, even Gradmann was criticized for ignoring French work on a phenomenon that Bloch believed was essentially European.[25] In brief, Bloch appeared to assess the adequacy of interpretations in this field by the success of their authors in capturing the changing social life, at both local and broader scales, and in avoiding interpretation by ethnic stereotypes.

The bulk of the works by geographers reviewed by Bloch was composed of regional studies, primarily in the form of doctoral theses, but also including books from some of the "collections" and shorter monographs and articles with a regional setting. In his early reviews he was quite supportive of these regional works and suggested that in France geographers (and more specifically "the immediate disciples of the great spirit who was Vidal de la Blache") had opened the way toward important areas of historical research and even of economic sociology.[26] However, he objected to the insistence by some authors that their work was specifically geographical. As Bloch explained, "do not the historian and the geographer meet in the same preoccupation for which human societies are the object?"[27] In another review, he identified their common object as "man in society," a term that more accurately depicted Bloch's approach, i.e. an approach which focussed not only on social phenomena in a strictly Durkheimian sense, but which also examined what Bloch described as the collective element in the individual. Despite noting their common concerns, Bloch admitted, "But in practice, it stands to reason that, in itself, the initial education received by this or that researcher cannot help but impose certain ways of describing and scrutinizing reality."[28] Bloch was not willing to grant the geographers a distinctive object of study, but did at least imply that their approach had its own characteristics.

In a number of his reviews, Bloch argued against the unspecialized regional monograph ("the monograph said to be complete"), writing that such works inevitably contained important omissions.[29] Prominent among the "gaps" which he found troublesome were social conditions (as in the role of particular social groups in the agricultural revolution), studies of family composition as linked to the "habitat," and more broadly, "currents of civilization" which affected regional development. He regretted that the questions of financing, agricultural technology, and *la base juridique* (the legal base) received so little treatment in works which claimed to study man-land relationships. Bloch also felt that the historical background was often insufficient for the explanations given. If a study was limited to a particular time period, the author, according to Bloch, should then admit to an inability to explain certain phenomena.

In an early review of Thérèse Sclafert, Bloch even went so far as to suggest that the regional monograph as practiced by geographers was ill-suited for the study of social phenomena as it tended to group together very distinct types of phenomena in a rather arbitrary way, ignoring others that should have been considered. As an alternative, Bloch proposed that regional studies be limited to particular social phenomena and cautioned that one must "define that nature of the phenomena studied with precision" and not "demarcate them too arbitrarily." These criticisms were very much like those of Simiand's 1910 review of the first Vidalian monographs, with the significant difference that Bloch did not completely dismiss the regional approach. Bloch added that the study of social phenomena with regional monographs was even more problematic in the past as demonstrated by Sclafert's study of medieval "Haute-Dauphiné" which overlooked the monetary system, the seigneurie, and *le facteur juridique*.[30]

Bloch also criticized the treatment of "geographical" phenomena in the regional monographs, arguing that they suffered from a lack of balance. According to Bloch, questions of physical geography and geology tended to be overemphasized, and those of the form and the distribution of fields were often ignored. He argued that at times the work could have been improved by a more careful use of maps and study of habitation.[31]

Despite his constant refrain of a lack of focus and of serious omissions in the contents of regional studies, Bloch did note by 1934 what he saw as some promising signs. Pierre Deffontaines' thesis, though not altogether satisfying, at least did not follow the old formula. It was, according to Bloch, "an overly hybrid work" failing to give a satisfactory treatment of either geography or history. He was nonetheless optimistic remarking, "a troubled science is a science in progress." On a more critical note, Bloch argued that this work illustrated a basic flaw in the conception of regional geography: "The unity of place is only disorder. Only the unity of the problem is central."[32] Bloch had more praise for works by Roger Dion and Henri Cavaillès which were effectively oriented around particular problems. Echoing his monograph on the Ile-de-France, he indicated that the problems which ought to be investigated were those of general interest. Bloch did eventually admit that such problems could not always be addressed directly in a work of geography, but argued that relevant problems should at least be formulated.[33]

In spite of all of his criticisms, Bloch found some of the geographical discussions in these regional works worthy of emulation. Daniel Faucher and Maurice Le Lannou, for example, were to be commended for putting their physical study "in its true place as an introduction and an analytic statement of relationships" and for highlighting those factors which actually contributed to the character of regional life.[34] More specifically, Bloch liked the treatment of climate given by Sion and Aimé Perpillou: Sion rightfully focussed on the light and the wind, "the two great Mediterranean realities," and Perpillou avoided the trap of relying on averages: "Is it not 'le temps qu'il fait' [the weather], in its often brutal integrality and reality, from which above all man suffers the repercussions?"[35] In a final example, despite some reservations over Théodore Lefebvre's thesis, Bloch did find much to praise in its treatment of geography. He liked Lefebvre's use of maps and illustrations: "By them the reader is constantly brought back to the terrain, that is to say to reality."[36] More broadly, Bloch praised his treatment of that reality as both concrete and as symptomatic of underlying social life:

A remarkably refined sense of the thousand links which link the man of the fields to nature, a great skill of seizing the value of symptom in visible things – to detect the climate or the hand of man behind the forms of vegetation, to recognize in the plan of a house the rural occupations which dictated it – these qualities, the most precious among those which an investigator must possess, ensure that the work never ceases to give a particularly reassuring impression: that of never losing sight of the concrete.[37]

Bloch's clearest overall statement on the character of a good regional monograph came in a very favorable 1941 review of Jean Despois' study of eastern Tunisia. According to Bloch, Despois focussed on a problem suggested by history, the succession and intermixing of contrasting *genres de vie*. He rightfully paid more attention to climate than to relief, made extensive use of "the study of the past," and looked beyond the boundaries of the area under investigation, studying those "forms of life" which came from elsewhere. As Bloch explained, "a sufficiently broad horizon" was, in fact, crucial to a problem-oriented study. In contrast to much work on *la vie juridique* (legal life), Despois examined links between the landscape and the law, noting, for example, the landscape changes which accompanied the changing forms of property. As a result, he gave, according to Bloch, "blood and flesh to controversies often a little deprived of contact with reality." As for the causal factor in this study, "it is only possible to discover a single factor of change: man himself."[38] In sum, although never as explicitly formulated as one might like, Bloch's view of the geographical contribution to a regional monograph appears to have been that of giving a sense of "reality" to the study of man rather than explaining that reality.[39]

Bloch appeared to dismiss the concept of landscape along with that of region as an appropriate object of study. This is demonstrated in his reviews of the landscape studies by the German geographers Gerda Bernhard and

Friederich Mager. According to Bloch, not only did such studies tend to get bogged down with detail, they also had important omissions such as the forms of the fields, economic circumstances, and once again, social structure.[40]

Nothing on the social structure – nothing, in truth, on man. I feel a few scruples at always repeating the same refrain. However, a "landscape of civilization" is only, after all, the expression of a society which, with "un élan vital" shapes and reshapes it in its own image. I admit to being unable to understand how one could explain it without studying this society itself. An empty shell rings hollow.[41]

Only "human societies" or "man in society" were to Bloch the real objects of such research, but as a means to that end, he did encourage a study of landscape.[42]

Although Bloch's advocacy of interdisciplinary research was far from unique, particularly at Strasbourg, the specific position which he adopted was his own. More critical of geographical works than his colleague Febvre, he did, nevertheless, think that geography had a contribution to make and was more willing than Halbwachs to take the geographers' works seriously.[43] In his many reviews of geographical writings, Bloch advocated an interdisciplinary approach drawing heavily on history, sociology, and geography. Arguing that such an approach was not in itself enough, he claimed that a truly effective work should also be problem-centered, and to him those "problems" of greatest interest were human ones, particularly those suggested by historical research. Studies of region, place, and landscape in and of themselves did not, according to Bloch, provide a satisfactory focus. One must look beyond them to questions relating to the common object of "human studies" – "man in society." Such questions could be addressed in a regional setting but only if one made comparisons to other regions. Even though geography often lacked explanatory power, Bloch urged geographers to raise the relevant questions, qualifying all the time any explanations proposed. To Bloch, geography's main contribution was in giving a sense of "reality" to such a study, in a depiction of the changing landscape and *le temps qu'il fait,* rather than in the particular problems which it might pose or explanations it might provide.

Bloch's view of geography expressed in these reviews was much like the sociologists' view of social morphology; he emphasized those elements of the "substratum" that were directly related to society and noted the intimate relationships between the "morphology" and "physiology." He also endorsed Simiand's criticisms of the regional approach. Both argued that comparisons were needed, that monographs should try to answer specific questions for large areas, and that geographical explanations were necessarily extremely limited. Nevertheless, his object of "man in society" was not equivalent to the social phenomena studied by the Durkheimians. Furthermore, Bloch did see some value in the use of a regional setting to investigate problems posed by history and argued that geographers had a contribution to make, going beyond localisation in space – to provide a sense of "reality" so necessary for meaningful work in "human studies." According to his metaphor, geography was to add flesh and blood to those studies, an emotive image hardly consistent with the efforts of the sociologists to create an objective science.

Although Bloch's initial role was that of a critic of geography, his own work on agrarian history soon involved him more directly in discussions within the field. As early as 1913, he had become interested in the question of field systems; at that time he praised the work of the English historian Frederick Seebohm on the open field system in England and suggested that similar work might be done for much of Western Europe. In Bloch's description of this "ancient agrarian system," he noted not only field patterns but also important "servitudes" such as those of cultivation and of passage (permitting others to cross one's land).[44] This question, redefined as that of the agrarian regime, was the major "problem" to which Bloch turned following his completion of *Les Rois thaumaturges*.

Bloch's treatment of the "agrarian regime" demonstrates how he approached what many saw as a geographical question. For Bloch, the problems raised by agrarian studies required an interdisciplinary approach and accordingly could be fruitful in the "understanding of man in society."[45] His writing on this topic contrasted with that done in France by both geographers and other historians, in part due to his rather sociological interpretation.[46] In both his definition of the regimes and his explanations of them, Bloch drew heavily on social factors. As in his essays on the transformation of serfdom, he drew on work which he had done on the rural populations of the Ile-de-France, but placed that work within a broader framework and focussed on a more specific issue. Although he was initially very critical of geographical explanations of the various agrarian regimes, with time he admitted that geographical factors had a role, albeit still a secondary one. He also eventually indicated that his initial statements on the role of the plow and on agrarian civilization should be reworked.

Thirteen years after his review of Seebohm, Bloch reviewed a work which was, he claimed, the first in French to address this issue of agrarian systems: a study of "Lower Belgium" by the historian Guilliame Des Marez. Bloch praised this work for its interdisciplinary approach, for its attempt to identify different stages of "the occupation of the soil," and for its use of *les plans parcellaires,* field maps which had been neglected by both geographers and historians. Although he also approved of Des Marez's attempt to criticize the German historian Meitzen, Bloch suggested that just as Meitzen had gone overboard with his ethnic explanations, Des Marez went overboard with geographical ones. Furthermore, Bloch argued that Des Marez had followed Meitzen too closely in his description of "agrarian systems," mixing together the categories of the distribution of *habitations* and agrarian *genres de vie* and using only the first as the criterion for his three systems. More generally,

Des Marez, like Meitzen, wrongly focussed on the material appearances.[47] According to Bloch, the focus should have been shifted to the "collective rules"; these rules not only marked the separation between different systems but were also often essential to the explanation of the material appearances themselves.[48]

The same year at the Société d'Histoire Moderne Bloch identified three agrarian systems in eighteenth-century France based on the form of enclosure and collective rights. In answer to criticisms that geographical conditions could be used to identify the different systems, Bloch argued once again that both the ethnic and the geographical explanations were too simplistic and that one could not ignore the very important role of tradition.[49] Bloch's approach here drew not only on the works of Gradmann (whose work on the habitat he liked so much), but even more clearly on those of English rural history. Bloch preferred an interdisciplinary approach which made a careful examination of the "terrain" but which also relied very heavily on a study of the changing "collectives rules." Only works on English rural history, it seems, had paid sufficient attention to the problems surrounding agricultural practices, common lands, and obligatory crop rotation, works which probably included those of Seebohm, Paul Vinogradoff, and Frederic Maitland.[50]

This research into agrarian "systems" or "regimes" occupied Bloch for the next few years, his first concern being to uncover the needed source material for France. In addition to the eighteenth-century investigations which he used in his paper for the Société d'Histoire Moderne, Bloch directed researchers' attention to a number of others including air photos and the forms of plows. In 1926, he published a research note in the *Annales de Géographie* on "le projet de code rural de Premier Empire" (his only publication in that journal) in an attempt to bring "this neglected but important source to the attention of geographers and historians of rural things, whose interests are about the same." Overlooked by such geographers as Sorre and Arbos, this rich source contained much of importance on agrarian practices.[51]

Another source which Bloch explored was that of *les plans parcellaires*. These maps indicated not only the field boundaries but also the location of individual parcels within them, subject to a particular form of cultivation and having a particular owner. As a result, they could illuminate both agrarian practices and changing property relationships. Using an analogy much like that adopted by the sociologists for social morphology and social physiology, Bloch argued:

Moreover, the study of the maps is, of course, not an end in itself. The material features that one sees inscribed there only have value because of what they reveal. They give the anatomy. What is important to us is the physiology of the living animal, I mean to say the rural community. But the anatomy is the initial understanding that the physiologist does not know how to do without, and reciprocally, it is only intelligible once its physiological side is scrutinized and described. Likewise, the "plan parcellaire" finds a place in the beginning as an investigative tool, one of the most practical and surest that there is; and at the end – once the little society of which the land is the carapace is well known and well understood – as the most immediately perceptible image of profound social realities."[52]

To Bloch, *les plans parcellaires* made this study of social phenomena concrete. One must never, he would argue, lose sight of such "perceptible images" when studying social reality.[53]

At the 1928 meeting in Oslo, Bloch gave a paper on the French agrarian systems in which he suggested that to "classify these systems, to retrace their evolution appears to be one of the most important objects that our studies can offer."[54] Speaking of the "technical rules," he argued that like language they were for the most part "social things." In veiled references to the work of Des Marez and Demangeon, Bloch argued that one must not make the mistake of only studying "exterior signs" of the agrarian customs and that furthermore, "it is advisable to separate, at least provisionally, the study of agrarian systems from that of 'habitat.' "[55]

By August, 1929, Bloch was immersed in his work on the agrarian regime as he had been asked to present a series of lectures at the Institute for Comparative Research in Human Culture in Oslo, for their newly established series on the "study of forms of development in the peasant community."[56] These lectures would form the basis of Bloch's next book, *Les Caractères originaux de l'histoire rurale française*. On August 9, Bloch wrote to Febvre, "I worked rather painfully on my lectures for Oslo. Obviously, in spite of so many obscurities and rash hypotheses, I will allow myself to make a book from them, because it is already something to formulate the problems . . . As you know, there is literally no synthetic work on the subject."[57] At this point, Bloch described agrarian systems as the collection of rules and tradition weighing on cultivators which dictated their methods of exploitation.[58]

Published by the Oslo institute in 1931, Bloch's book was the first of three French works to deal with the agrarian regimes in France. The other two were *Histoire de la campagne française* (1932) by the historian Gaston Roupnel, and *Essai sur la formation du paysage rural français* (1934) by the geographer Roger Dion.[59] None of these works were strictly limited to a study of the agrarian regimes, but here the discussion will focus on this topic which most clearly illustrates their differing approaches.

Although all three authors were concerned with agrarian systems, they perceived and interpreted them differently. Of the three, Bloch's approach was by far the most sociological. For him, a close examination of social organization was required. For example, he argued, "An agrarian regime is not only characterized by crop rotation. Each of them forms a complex network of technical formulas and of principles of social organization."[60] Throughout his discussion he stressed the obligatory character of the social rules using such terms as "obligatory," "forced," "prohibited," "constrained," "collective pressure," and "imperious will of the group." The agrarian regimes, as he now called them, were social institutions not only in their characters but also in their causes. For example in his description of the open field system in northern France he remarked, "beneath these perceptible features, let us know how to see human causes. Such a regime could only have arisen thanks to a large social cohesion and a fundamentally communitarian mentality."[61] To Bloch, both geographical and ethnic explanations of the regimes were inadequate. Instead, he preferred to associate them with differing types of "civilizations." Even more than Marcel Mauss who had helped to popularize this concept, Bloch attempted to separate the concept of "civilization" from that of "geographical unities" arguing that the agrarian phenomena could not be enclosed within geographical boundaries.[62] Bloch also drew analogies between agrarian and linguistic phenomena and suggested that one ought to consider the "religious factor" in the constitution of *les terroirs,* thus associating his concept with two of the key social factors for the Durkheimians.

Stressing differences in collective practices, he identified three different agrarian regimes for France: an open field regime where enclosures were banned, fields followed a regular elongated pattern, and a strict system of crop rotation was practiced; a second regime also without enclosures, characterized by irregular fields and less rigorous collective constraints which weakened somewhat more quickly; and that of the enclosed fields which was more individualistic, with the exception of the common grazing lands which could be subject to strict collective constraints. He also added that systems of crop rotation varied between these regimes with a triennial system associated, with some exceptions, with elongated fields and a biennial system with open and irregular fields.[63]

As for the relationships between Bloch's regimes and geographical factors, he noted that the region of open and elongated fields tended to be found in the north, that of the open and irregular fields in the south, and that of the enclosed fields in the west, but he quickly added many exceptions and stressed the complexity. Although shying away from correlations between his regimes and characteristics of the physical geography, he did remark that the region of the enclosures tended to be associated with regions that were "often hilly, in any case with thin soil," and also indicated that he would not consider the "peculiarities of certain areas controlled by very particular natural conditions," such as mountainous regions.[64] Although he insisted on a close examination of the terrain, Bloch was both hesitant to endorse any explanations that appeared geographical and yet not ready to totally rule out a place for geographical factors. In one of his most explicit statements he wrote that "one can, nevertheless, distinguish in France three great types of agrarian civilizations, closely linked to both the natural conditions and human history."[65] Despite this mild concession to natural conditions, Bloch refused, for example, to associate the biennial rotation simply with the dry summers of the Midi, claiming that it was more clearly linked to a contrast between civilizations.

In addition to stressing the social characteristics associated with the various agrarian regimes, Bloch also noted one technical factor, that of the type of plow used. *L'araire* was associated with the system of open and irregular fields, and a plow mounted on wheels, *la charrue*, was associated with the divisions of fields into long narrow bands. Bloch noted that *la charrue* required considerable space in which to turn, a technical and legal problem which could be lessened if one cultivated the fields in the long bands associated with the open field system. Despite some later accusations, however, Bloch did not give a simple argument of technological determinism, which in any case would have been very much out of character.[66] He did comment that the contrast between the layout of the fields in the regime of elongated fields and that in the regime of open and irregular fields "in all appearances comes down to an antithesis of two techniques," but he also argued that *la charrue* by itself could not account for the first field system since it required long fields but not necessarily narrow ones. The system could only be explained by "ideas, so profoundly anchored in the peasant consciousness" regarding the division of the land, and also presupposed strong "communitarian customs" allowing the adoption of such a system. In the end, begging the question, Bloch wrote, "Thus, less ambitiously, let us limit ourselves to noting that as far back as we can go, *la charrue*, mother of elongated fields, and the practice of a strong collective life join together to characterize one very clear type of agrarian civilization; and the absence of these two criteria, a completely different type."[67]

In contrast to Bloch, Roupnel and Dion gave far greater weight to geographical factors, though in somewhat differing ways.[68] Roupnel, for example, viewed the northern agrarian regime of open fields as one particularly favored by the land and climate, related in part to the races which occupied the region, and perfected by Christian faith and practice. Also in contrast to Bloch, he saw the agrarian regimes of the south and west not as products of separate civilizations but rather as gradual extensions of a northern system. That slow extension and the differences in physical geography were, he suggested, enough to explain the contrasts; he objected to Bloch's intimation that the varying forms of plows might have had a role to play. Since his argument was one of diffusion, he made little attempt to specify the regions associated with the differing practices, resting content with occasional references to the physical geography.

Though a geographer by training, Dion was somewhat more open than Roupnel to social interpretations, attacking those based on natural regions and pushing for an examination of the relevant customs and constraints. Nevertheless, his concerns were more directly those of landscape than of the broader social relationships which formed the focus of Bloch's book and of which the agrarian regime remained only a part. Paying less attention than

Bloch to historical change as he jumped from Tacitus to the eighteenth century, Dion was more willing to draw on physical geography including the climate and soil type in some his explanations. At the same time, however, he attacked "this geography based on geology which many of our contemporaries praised for having provided the definitive exploration of the rural landscapes."[69]

Dion's identification and explanation of the agrarian regimes also differed. In contrast to Bloch, Dion identified only two agrarian regimes, the open fields of the north and the enclosures or *le bocage* of the south and west – though in his regional definition, much like Bloch, he stressed the complexity and noted many exceptions. Whereas Bloch saw the open fields of the south as indicative of a separate civilization, Dion viewed them as an exception, this despite an association noted with earlier settlement. To Dion, perhaps, such civilizations were meant to characterize the eighteenth century, whereas Bloch interpreted civilizations as having greater historical depth.[70] Quicker than Bloch to adopt ethnic explanations, Dion suggested that the communitarian character of the north stemmed from the pastoral life out of which the regime grew, which had Germanic origins, and that "the agrarian liberty" of the south and west had Roman origins. Furthermore, Dion saw the agarian liberty of the south as much more pervasive than Bloch.[71] Also much like Roupnel, but unlike Bloch, Dion treated the field systems in conjunction with those of habitat, and argued that *la charrue* did not coincide neatly with the area of the southern system.

These three books sparked a long debate in the geographical literature, a debate in which Bloch played an active part. In general the geographers praised Bloch's work but felt that he had paid too little attention to the relevant geographical factors. Demangeon argued that he should have paid more attention to the nature of the soils and the kind of plants cultivated. Also, as the leader of habitat studies, he claimed that more was needed on the character of rural settlement, noting the extent of dispersion and the places chosen. This was in opposition to Bloch's view that studies of the habitat and the agrarian regime should in the first instance be separated.[72]

Sion was more appreciative of Bloch's social orientation but even he wrote somewhat critically, "But when must one manipulate the large keys of historical materialism, more or less retouched by geography, and when the thin elegantly serrated key, made fashionable by sociology? In truth, they suit different temperaments. And Mr. Bloch, while understanding and demonstrating the imporance of the first, seems sometimes inclined to discount their use a little."[73] As examples, Sion argued that the arrangement of fields in the northern regime made common grazing land the only practical solution; that Bloch should have looked more closely at the soil as an explanatory factor for the lay-out of fields, especially in the south; and that to explain *le bocage* one should note the humidity favoring the growth of pasture. As for Bloch's treatment of social factors, Sion remarked that the concept of "civilization" was "still a somewhat vague notion, but the very effort that will be needed to clarify it will be fruitful." In addition, he suggested that a study of its adaptation to the physical, economic, and social milieux might be helpful, i.e that civilization was not purely a social phenomenon.[74]

Despite their reservations, the geographers appeared to prefer the scholarship and caution that marked Bloch's book to the works of Roupnel and Dion, even though these latter paid greater attention to geographical factors. Roupnel, for example, was criticized by Demangeon and Faucher for arguing by affirmation and for giving an unconvincing treatment of the agrarian civilizations. Dion, though closer to representing the "scientific" school of geography, was attacked in public by his supervisor Demangeon for his "risky hypotheses" and also criticized by Sion who suggested that Bloch's three-regime argument gave a more convincing portrayal of important technical and social differences than Dion's two-regime one. Furthermore, neither Demangeon nor Sion liked Dion's ethnic explanations.[75]

Bloch soon entered the debate. Although he did not review Roupnel's book, he told Febvre (who wrote the review for the *Annales*) that he lacked Febvre's indulgence toward the work. Reflecting his early training, Bloch explained: "It is not a question of being vigorous like Roupnel or distinguished like Halphen; it is a question of being exact, which means at the same time close to life."[76] The reference to Louis Halphen was explained earlier in his letter. He found Halphen's work on the invasions intelligent, well informed, and clear, but at the same time "far from life."[77] Its literary style masked the "concrete" and the "different," a fault which Bloch attributed in part to a lack of geographical background. He continued: "Even though de Martonne is not always amusing; even though Demangeon's human geography is sometimes very inconsistent; let us carefully protect our system of historical-geographical training!"[78] Bloch turned to geography as a way of getting close to life but at the same time, like the "positivist" historians who preceded him, demanded careful documentation; together, a sensitivity to "life," which implied the "concrete" and the "different," and a careful examination of the evidence could help one to be "exact."

In spite of the many reviews of Dion's work, Bloch wrote a review for the *Annales*. As he explained to Febvre in early 1936, "Dion has already been abundantly reviewed and criticized from right and left; but mostly by geographers who have not really followed, to my mind, this big chink in the armor: which is that it is, in fact, very difficult for a geographer to think in truth in terms of duration. Rest assured, I did not say that."[79] In his review, Bloch was less direct, simply indicating that more was needed on the origins and development of an agrarian regime. More specifically, Bloch criticized Dion for downplaying the strength of collective ties in *le bocage* and for his recognition of only two agrarian regimes. This question was not simply scholastic, he argued, since it implied that differing origins and different problems were associated with each.[80]

With time some of Bloch's positions on agrarian regimes began to change. In a 1936 review of Daniel Faucher's articles, he admitted that physical geography could help to explain some of the characteristics of the southern regime, an indirect answer to some of the criticisms of Demangeon and Sion. However, he stressed that the geographical problem must not be posed too narrowly. The problem of the "North" and the "Midi" should not be seen as simply one of the agrarian geography of France, but instead of "its entire social geography." [81] As such, it would map not only differences in "field patterns" but also in law and custom, in language, and even in house styles. This many-sided approach should help to avoid the dangers stemming from the "obsession with prehistory or the ethnic factor."[82] In a very critical review of Paul Fénélon's work on Périgord (1942), Bloch indicated that his initial position requiring a separation of the study of the "habitat" from that of the agrarian regime had changed. Bloch now conceded that the two might be studied together but still differed from Demangeon in stressing the links of both problems with that of social structure. In other changes of position, Bloch eventually admitted that the term "agrarian civilization" was unclear and that his "thesis on *la charrue*" needed to be reworked.[83]

Bloch's approach, which was interdisciplinary and in which both sociology and geography had key roles to play, set his work apart not only from works in geography but also from those in history. Some found his interpretation too sweeping. Lot suggested, for example, that "communitarian spirit" was not as old as Bloch implied and Michel Augé-Laribé also charged that Bloch had exaggerated the continuity between the past and the present.[84] On a more positive note, though indicating some unease with Bloch's social orientation, Febvre noted that the originality of *Les Caractères originaux* stemmed from the range of disciplines on which it drew, a happy contrast to the earlier works of Sée and Augé-Laribé. As a result, Bloch's book served the needs of "historians concerned with realities" and "geographers curious about origins."[85] To André Deléage, the greatest merit of Bloch's study was its attempt "constantly to bring together sociology and agrarian history" and its greatest innovation, the attention given to *le terroir*.[86] In England, Sir John Clapham hailed the book as long overdue and praised Bloch's comparative approach and his work on the *plans parcellaires*. Richard Henry Tawney, writing for the *Economic History Review* said that Bloch's work had the particular merits of drawing effectively on regional and local studies and of showing the possibilities of comparative economic history.[87]

Given the role which sociological concepts played in Bloch's book, it is not surprising that the sociologists also expressed interest in the work. Shortly after *Les Caractères originaux* appeared, Bloch spoke before the Institut Français de Sociologie on "the problem of the agrarian regimes." This institute, founded by Mauss and dominated by Heilbron's *chercheurs*, welcomed practitioners from many fields – illustrating both a continued interest in sociology by those outside the discipline and its continued dependence on other fields.[88]

Clarifying what was now his preference for the term "agrarian regime," Bloch said of the alternative expression, "agrarian system," "I find it a little too rigid and I prefer a less compromising word, such as that of regime."[89] Discounting geographical factors, he preferred social ones and even cautiously suggested that "phenomena of a religious order" could help explain the shape of fields in the south. Referring to his technical explanation relating to the form of the plow, Bloch downplayed its significance; he claimed here that such factors were clearly secondary to the social forces and "the fundamental difference is essentially a difference of social mentality."[90]

In the discussion that followed, Simiand mentioned that "the technical factors appear to me to have been and to still be more conditioned than conditioning."[91] To him, Bloch's explanations in terms of "social and religious factors" seemed to be those most worth pursuing. Drawing on the work of Henri Hubert and his own interest in the concept of "civilization," Mauss argued that one could relate three movements within the early Celtic civilization to the agrarian regimes and that too often contributions which came from this Celtic civilization were attributed to the Greeks and Romans. Bloch noted that he had been in correspondence with Hubert. However, here as elsewhere he hesitated to adopt ethnic labels for his civilizations.[92]

Deeply influenced by the sociologists, Bloch's work retained its own character.[93] He remained a staunch advocate of the comparative method but preferred his own particular version of that method which highlighted the "original characters" and, as always, he hesitated to formulate broad generalizations. The "original characters" which formed the central theme of his study were those of the contrasts found within France, as well as the slowness of agricultural change. To Vidal, France's particular character was based on the diversity and complementarity of its geographical regions; Bloch, on the other hand, attributed its character to contrasting civilizations, a concept associated in part with the sociologists, which Bloch endorsed but failed to clarify.

By the early 1920s in discussions of Febvre's *La Terre et l'évolution humaine* and of the *Géographie Universelle*, it became clear that Bloch, more than his fellow social historians, Febvre and Hauser, sided with the sociologists rather than geographers. He attacked the regional approach, pushed for a comparative investigation of particular problems, and emphasized the need to examine social factors and historical change. Even Sée, who agreed that history had more to learn from sociology than geography, relied in his own work on legal definitions rather than studying the "collective representations" which Bloch argued were crucial. Later, Bloch clarified his position on geography in his numerous reviews of the inter-war geographical works. Geographical research, though useful, was seen as too one-sided and needed to be incorporated into an interdisciplinary approach in which both the study of *le terrain* and of collective representations gave the work a sense of "reality." Geography's object of study should be that common to all the other "human studies" – social man or man in society; this object should be approached through the investigation of various "problems," problems which to Bloch appeared to be raised most effectively by history. Unlike Simiand, he did not dismiss human geography as a discipline. He did, however, relegate it to a truly modest role.

His approach became clearer in his own work on the agrarian regime, a topic suggested not by work of the French sociologists or geographers but of the English historians. Although stressing "collective rules" highlighted by the likes of Seebohm, Vinogradoff, and Maitland, Bloch also made a careful examination of *les plans parcellaires* and looked for regional variations. These variations, Bloch insisted, could only be explained by reference to differing agrarian civilizations, an argument which even Sion found to be overly sociological, giving too little room for geographical explanations. Although ending with a sociological argument, Bloch was reluctant to pursue it too far away from the "intermediate" civilizations which he knew best. More reluctant than Hubert or Mauss to label his civilizations, Bloch left this question to the "mists of prehistory" which he, himself, so often belittled.[94]

9

An expanding view: Marc Bloch's later projects

Following Marc Bloch's move to Paris in 1936, much of his time was taken up by projects begun sometime before. It was a time of multiple commitments and unfinished business. Not only had he promised several volumes to Berr's series, "L'Evolution de l'Humanité," he was also now responsible for directing and contributing to his own series, "Le Paysan et la Terre." His new position in the chair of Economic History at the Sorbonne further restricted the kinds of work he could undertake. As he pointed out to Febvre, he would be forced at the Sorbonne to leave aside interests that did not fit neatly under the label of social history. This would include some of his more politically oriented work such as that on the *rois thaumaturges* and on *l'idée de l'Empire*. As he confided to Febvre, "One of my preoccupations is to see how I will be able, on occasion, to reintroduce these things; and for the time being I certainly won't speak of this to others than you."[1] The disruption of the Second World War meant that several of these later projects would remain unfinished. Nevertheless, during the many idle moments which the war also brought,[2] Bloch had time to reflect and write on the nature of history, grappling yet again with issues in the pre-war debates between historians, sociologists, and geographers – in a work which would also never be completed.

In these later works, Bloch was moving towards a wider perspective in which he dealt with longer periods of time, larger areas, and even non-European civilizations.[3] In part, this was an outgrowth of the synthetic works which he had agreed to write as a mature scholar, but it was also a logical development of his own interests and initiatives. Still committed to problem-oriented research, he began to pose his problems in yet more general terms and to make his work even more sociological. At the same time, he clarified how his approach differed from Durkheimian sociology, a difference due in part to his continued attraction to a geographical perspective. These changes are evident in Bloch's writings on economic history (many of which stemmed from his research for Berr's series), his book on feudal society (also for Berr), his contributions to his own series, and his unfinished work on the nature of history.

As Bloch worked on *Les Caractères originaux* in 1929, he was already concerned about his long-overdue commitments to Berr's series, and on completing the manuscript in 1930 wrote to Pirenne, "I ought to get to work actively on books for Berr's series."[4] In 1924, Berr had asked Bloch to take on volume 44, which had been attributed to Georges Bourgin and was then entitled "Développement économique: vie rurale et vie urbaine." One of the volumes in the section on the origins of Christianity and the Medieval Ages, volume 44, was to have been followed by a volume entitled "Le Commerce maritime: les sociétés marchandes" (volume 45) by Prosper Boissonnade.

Bloch's responses to Berr's request demonstrated that in his view economic history was very broadly conceived and was in fact a form of social history. He worried about the separation between his volume and Boissonnade's, wondering who would treat the issue of the origins of capitalism and how he could separate the economic development of towns from merchant societies. In addition, he noted that he could not draw a strict separation between "economy" and the *vie juridique (et même vie morale)*. This would cause a problem with the division between his volume and those of Ferdinand Lot on the end of the Carolingian Empire and the feudal regime and of Edmund Meynial on medieval public power and law. Bloch's desire was to write a wide-ranging comparative economic history of the European West – a task he thought would take at least two volumes. In them, he hoped to treat more than a thousand years of "economic evolution or rather social evolution as viewed through the economic aspect."[5]   After Boissonnade's death in 1935 Bloch was listed for both volumes 44 and 45 which were now entitled: "Les Origines de l'économie européenne (VIᵉ–XIIᵉ)" and "De l'Economie urbaine et seigneuriale au capitalisme financier (XIIᵉ–XVᵉ)."[6]

By 1933, however, yet another volume had become the top priority – a volume even more in line with Bloch's social perspective. He agreed to replace Lot as author for the volume on feudalism, and Berr was keen to have that appear as soon as possible. As he began work on this new book, Bloch admitted that his work on the economic histories had been difficult because of the need to limit the discussion on social structure. In this new book, that would no longer be a problem since this work would be, he suggested, "above all, a study of social structure."[7] Despite the pressure from Berr, the manuscript, *La Société feodale*, was not completed until 1938.

As late as 1942, he toyed with the idea of returning to his books on the origins of the European economy, but without much enthusiasm since he did not like the constraints of working within a series, and, as he explained to Febvre, "I feel more 'rural' and monetary." He now saw the interesting economic question more in terms of "the history of French currency."[8] Some fragments from his proposed work on the origins of the European economy do remain, however, and demonstrate that Bloch continued to view economic history as a form of social history.[9]

Bloch addressed the theme of the dependence of technology on social structure in two 1935 articles, on medieval inventions and on the water mill, that had been written originally for his book on the origins of the European economy.[10] Whereas Commandant Richard Lefebvre des Noettes argued that slavery declined because of the invention of modern harnessing techniques, Bloch suggested that the opposite had been the case – a given invention would only be adopted if it served the needs of the society in question. In his article on the water mill, Bloch examined the invention and use of this form of technology – which interestingly enough was also highlighted by Simiand in his critical review of the Vidalians. Simiand had used this example to illustrate the limits of physical explanation;[11] Bloch used it as another example of the dependency of technology on social structure. Present since ancient times, the water mill did not become widespread until the medieval period when labor was scarcer. Bloch, however, was not willing to reduce the relationships between society and technology to a formula. As he wrote in a later article: "But it [the lesson of the evolution of techniques] also reveals that there are no privileged causal wave trains; no phenomena that are always and everywhere determinant opposed to perpetual epiphenomena; that on the contrary any society, like any spirit, stems from constant interactions. The true realism in history is to know that human reality is multiple."[12]

Two articles published posthumously under the title of "Une Mise au point: Les invasions," can also show how Bloch addressed the question of the origins of the European economy. According to Perrin, these articles were taken from Bloch's 1941–1942 course at Montpellier on the Barbarian Invasions. In any case these pieces appear to be a study of the "social substratum" out of which the European economy grew.[13] Bloch sketched *l'occupation du sol, le peuplement,* and the role of the invasions in forming the "European economic milieu." In them, Bloch relied heavily on the still somewhat nebulous concept of "civilization" and observed that differences in technology could not account for the contrasting economic structures of the Roman Empire and Germany. As he explained, "The reasons for the striking antitheses between the two economies lies elsewhere; they come down to an antithesis between two social structures."[14]

In Bloch's various economic writings, he challenged some of Simiand's positions on the study of economic phenomena, but with time he came somewhat closer to the type of approach Simiand had advocated. Although Bloch shared with Simiand a belief that economics must be treated as a social study, their approaches to the questions of technology differed. In 1903, Simiand had cautioned against an examination of questions of technology and legal history in studies of economic history. By contrast, Bloch gave a very large place to their study discussing not only the watermill but also land ownership and the legal restrictions associated with agriculture and personal status.[15] In other work, Bloch began to tackle the general questions on the character of economies and on the nature of money to which Simiand had pointed in 1903. Bloch argued that the distinction often made between natural economies and monetary economies was misleading and that the term "closed economy" was often misused. According to Bloch, money should be treated as a social phenomenon. He wrote, for example, "all research on payments must become 'social,' if it wants to reach its object."[16] To Bloch, this entailed an examination of divisions within the society, investigating which groups stood to benefit from changes in the value of the currency.[17]

Bloch's plans for a history of French money marked a step even closer to Simiand's way of thinking.[18] Less attention would have been paid to origins and technology than in his proposed work for Berr and more to the general economic phenomena of a given place over time, without, however, strictly limiting his study to France. He intended to examine the changing methods of payment, of measuring values of exchange, and of conserving value for future needs. His work, which was not written, was to be one of synthesis and to focus on very general changes in monetary systems and the crises which developed. As François Dosse has argued, Bloch was not alone in his interest in such questions. During the 1930s when economic crises and issues were so evident, there was considerable interest in economic questions as demonstrated by the publication of works on prices by not only Simiand (1932) but also Ernest Labrousse (1933) and Henri Hauser (1936).[19]

When Bloch turned from questions of economic history to questions of the character of societies, he also adopted an increasingly sociological approach. This was well demonstrated in *La Société féodale*, the last major work he would complete. Nevertheless, the work was still clearly one of history not sociology. The problem that Bloch chose to address, though very much a social one, was stated in historical terms. He proposed an exploration of the "originality" of feudal society explaining, "a book on feudal society can be defined as an effort to answer a question posed by its very title: by which peculiarities has this fragment of the past merited being separated from its neighbors? In other terms, it is the analysis and explanation of a social structure, with its connections, that one intends to attempt here."[20] Despite the breadth of his topic and the length of his work, he wrote to Febvre, "in a word, it is a question of a single book, in two volumes and accordingly, one problem." Nevertheless, Bloch refused to define "feudal society" at the outset, arguing that such a discussion should come at the end of the study.[21] The introduction was devoted instead to an examination of the changing meanings of the term "feudalism."

Western and Central Europe were chosen as the framework for his study both because that was the location of the first feudalism to be so labelled and because within it one could find "a tonality of common civilization." He referred to that region simply as "Europe" since that was the original use of the word in the Middle Ages. In contrast to the Ile-de-France, Europe, thus defined, had a meaningful social content, and in contrast to *Les Caractères originaux*, there was no longer any pretense of using political boundaries as a framework. Instead the region fitted his criteria for *les cadres* of social geography, whose boundaries, he had argued, could be identified by the proper use of the comparative method.[22]

In an attempt to get to the heart of feudalism, Bloch chose as his "vital lead" the social tie between a vassal and his lord. He proposed to examine that tie in its most fully developed form as found in "the heart of the ancient Carolingian Empire" during the tenth to the twelfth centuries.[23] Here Bloch seemed to have kept in mind both Simiand's early attack on the chronological idol, which advocated an examination of *le type normal,* and Halbwachs' advocacy of an examination of the most characteristic period of the healing ritual, instead of a tracing of its rise and fall. For his part, Bloch explained his choice of an approach: "doesn't it [feudal nomenclature] offer us a reliable mirror of a social regime which, though heavily marked by a particularly composite past, was nevertheless above all the result of original conditions of the moment? 'Men,' says the Arab proverb, 'resemble their own time more than their fathers.'"[24] In contrast to his discussion of "agrarian regimes" in *Les Caractères originaux*, he argued that the specific origins of the social tie between a vassal and his lord were not only obscure but also not very significant.[25] Arguing that the period was one in which formalism was extremely important, Bloch claimed that to understand the bond between a vassal and his lord, one must examine the rite by which it was established, the rite of homage.

Bloch's approach here fell between those typical within the historical establishment and those advocated by the sociologists. Despite his focus on the most characteristic period and social tie, he still gave a more chronological treatment than some of the sociologists would have thought appropriate. The design of the series may well have contributed both to Bloch's failure to examine long-term origins and his attempt to trace "the very curve of the institution."[26] The individual volumes of "L'Evolution de l'Humanité" were intended, according to Berr, as "partial syntheses within the complete synthesis." Not all the books successfully made such links, but they remained one of the goals. Bloch's task was further complicated by earlier plans made when Lot was to be the author and the volume was entitled "La Dissolution de l'Empire Carolingien et le régime féodal." As Lot later noted, Bloch could not effectively explore some of the more long-term roots of feudalism since they lay outside his assigned period.[27] Even so, his picture remained more schematic than demanded by these restrictions, as he paid little attention to such contradictory elements as the growth of the bourgeoisie as a social class.[28] His work was marked by a sociological interest in the central institutions and ritual, even if those interests were explored using what remained essentially an historical approach.

Although his subject was defined as the "originality" of European feudal society, Bloch ended his book with a discussion on feudalism as a "social type," a discussion which he himself noted was not central to his book. There he spoke of stages of evolution and tried both to identify the salient features of European feudalism and to suggest comparisons to elsewhere.[29] However, because of his focus on "originality," the concluding section is not totally convincing. It stands somewhere between a characterization of an ideal type and that of a particular historical society, as he argued that the European case could not be divorced from its particular history. Having concluded that the bond of vassalage characterized the "originality" of a specifically European feudalism, he also suggested that one might speak of a Japanese feudalism since there vassalage, albeit in a different form, was very important. Although this approach was consistent with his belief that historical concepts should not be divorced from their historical base, it led to a somewhat confusing discussion of social type.[30]

An interesting contrast can be made between Bloch's treatment of feudalism and that of Marcel Granet, his former colleague, who was more closely associated with Durkheimian sociology. In 1936, Granet gave a series of lectures at the Institute for Comparative Research in Human Culture, in Oslo, on Chinese feudalism. Whereas Bloch gave a very schematic picture of feudalism allowing only a minor role for contradictory elements, Granet saw feudalism as a "regime of transition," of interest to sociology because of its "genetic" implications, and as a result gave a much greater place to elements which ran against its character. Even though less bound by a schematic type, Granet attempted to use the Chinese case to explore the character of feudalism in general – a contrast both to Bloch's focus on "originality" and his belief that there was little to justify the application of the word "feudal" to Chinese society. More interested in evolution and generalities, Granet was less concerned with specifically identifying feudalism as a social type.[31]

Although geographical discussions were not central to *La Société féodale*, Bloch did make some use of geographical concepts and detail. Following a common practice among historians, Bloch began his book with a discussion of the milieu, which he labelled as "the social milieu." Given his strong social orientation, this section, though not ignoring geography, stressed the atmosphere of disruption caused by the invasions and the character of communications, trade, travel, and the *mentalité*.

Bloch's opening discussion of the invasions forms an interesting contrast to Louis Halphen's work *Les Barbares* which Bloch had criticized for being too detached from life in part because of its lack of sensitivity to geography.[32] (Then a professor of the auxiliary sciences of history at Bordeaux, Halphen later taught at the Ecole Pratique des Hautes Etudes and in 1936 replaced Lot at the Sorbonne.) Despite Bloch's accusations, Halphen's book did not ignore geography and was full of what could be viewed as geographical references.

Page after page was cluttered with place names. By contrast, Bloch used fewer place names, preferring instead references to regions. Where possible he relied on names of the times, such as "France Orientale" for Germany and "les Solitudes" for the Tisa plains and the middle Danube. In addition he gave much more weight to the characters of the habitat, *les genres de vie*, the terrain, and the climate. In part, this contrast between the approaches of Bloch and Halphen reflects that between the old historical geography, on which Halphen relied to chart the progress of the invasions, and Vidalian geography, used by Bloch but adapted for his own purpose to become both more social and historical.

In addition, an important difference between the two works lay in Bloch's stress on contemporary perceptions. Using quotations and anecdotes, he attempted to demonstrate just how different the place and time was from those that would be familiar to his readers. As an example, his discussion opened with the following quotation from the bishops of Reims in 909, "You see exploding before you the anger of the Lord . . . There are only depopulated towns, monasteries thrown down or burnt, fields reduced to solitudes . . . Everywhere the powerful oppresses the weak and men are like the fish of the sea that devour each other pell-mell."[33] Halphen, more in line with the *historiens positivists* excluded such anecdotes, something which Bloch felt made his book both boring and imprecise. As Bloch wrote to Febvre, "Problems to resolve: keep the maximum of proper names compatible with a 'synthetic' exposition. But as one must not be too long and must avoid, above all, the reproach of relying on 'color' or 'anecdotes,' one eliminates implacably all human detail. One forgets that a name without content means nothing."[34] The content which Bloch felt needed to be associated with the names, be they names of places or of people, was that which illuminated the "different."

The remaining sections of his discussion on the milieu were devoted to what he termed "the conditions of life and the mental atmosphere." Bloch included some discussion of demography and habitat, linking them together as the advocates of social morphology had suggested, and he devoted somewhat more space to communications, travel, and trade. Thus, Bloch followed his own earlier intimations that although the material substratum was of interest, it was the life of the society that was most important.[35] Given his social orientation, Bloch spent even more time, in his section on the "social milieu," discussing the *mentalité,* collective memory, and the "formalist" character of law and custom. There, although making observations on the contribution of an untamed nature and poor nutrition to a certain "roughness" and "emotional instability," Bloch focussed on such topics as the dualism between the daily languages and Latin, and contrasts between *mentalités* of different social groups. Paying particular attention to the operations of collective memory, he looked at the role of both the written record and the epics, noted errors in the memory, and discussed the very fluid character of "formalist" law, i.e. he looked at precisely those factors which he had earlier accused Halbwachs of ignoring.

Unfortunately, *La Société féodale* was not reviewed in the Durkheimian journals.[36] The historians who reviewed Bloch's work had significant reservations. Febvre, for example, argued for a form of history that was far more traditional than that practiced by Bloch in *La Société féodale*. He wanted a more complete and less schematic picture with greater attention to individuals and to the character of *le pays* and the landscape as well as more precision in the time periods. The book, he charged, was too abstract because of an overly sociological approach. Others too were uncomfortable with the sociological character of this work. Although praising its originality, richness, and comparative approach, Perrin also charged that the work was too schematic and abstract and paid too little attention to individuals. Lot suggested that Bloch's approach was overly analytic. He seemed uneasy with Bloch's discussion of feudalism as a type (preferring a stricter limitation to the European context) and advocated a more legal approach. The legal historian Paul Ourliac, writing for the *Bibliothèque de l'Ecole des Chartes*, was even more critical. He objected to Bloch's attempt to characterize feudalism by personal ties of dependence and as such belittle the significance of the "fief," and he disliked the emphasis Bloch gave to rites. Although more complimentary, the English historian F. M. Powicke also suggested that the work was not perfectly balanced. According to his fellow historians, Bloch had clearly moved too close to a sociological approach.[37]

In Bloch's writings for Berr and related work, his approach was closer than ever to Durkheimian sociology as he explored very broad social questions. In accordance with Simiand's 1903 proposals, his economic history of Europe was to pay considerable attention to prices and money, and not unlike the positions of both Simiand and Halbwachs, he was very interested in the more fundamental questions related to the nature of economies. He was more interested than Simiand in relating economic development to internal divisions in society and more reluctant to establish general relationships and laws and to exclude technology and legal history from study. Increasingly, however, he turned in the direction to which Simiand had pointed, focussing on money, rather than technology and origins.

In *La Société féodale*, Bloch focussed on the general characteristics of feudal society, made wide-ranging comparisons, and attempted to treat feudalism as a social type defined by its fullest development. Although for many historians, his approach had become too sociological, his work remained one of social history rather than sociology. It focussed on the "originality" of European feudalism, even in its discussion of social types, and it traced the growth, decline, and survivals of European feudalism, even if avoiding the debate over origins. Here, Bloch addressed a very ambitious topic, encompassing Europe as opposed to just France, and even attempted some limited comparisons to elsewhere. Despite his widening lens, however, Bloch's interests were still focussed on the Ile-de-France.

Bloch did not ignore the milieu in his writings for Berr, but chose to treat it as a social milieu. He used the terms "social substratum" and "milieu" rather loosely as he did not make a clear separation between them and social life. Drawing more heavily on the concepts of Vidalian geography than those of the earlier historical geography, he rejected, nevertheless, the Vidalian use of natural regions. Instead he turned to what he called "the social geography" both for the general framework for his writings (a socially defined Europe) and for the comparisons which he made within it. The works of geographers were used to clarify the discussion and to add some life and color, but Bloch's writings for Berr were neither inspired nor defined by geography.

Bloch's use of sociology and geography during his later years was also evident in his direction of his own series, "Le Paysan et la Terre." In November, 1934, he received a letter from Gaston Gallimard, proposing that Bloch direct a new series. There, Gallimard wrote, "We would be happy, at the publishing house of N.R.F., to undertake the publication of about ten important works that could serve to constitute a history of the most important periods of the peasantry in France and in Europe."[38] To launch the series, Gallimard suggested that Bloch write his own book bringing together his dispersed writings on serfdom. After satisfying himself that they were willing to publish a serious series, Bloch agreed, seeing the series as "an interesting work of intellectual propaganda."[39]

In the course of the negotiations between Bloch and the Librairie Gallimard, the form of the series began to take shape. Initially, Bloch appears to have insisted on "a necessary suppleness," and he also argued that the books not take the form of manuals.[40] Shortly afterwards, Gallimard wrote to Bloch to verify the agreement reached: "Naturally, it is not a question of composing a continuous history of the peasantry in France and in the world, but to devote a volume to each of the important moments of this history, without contemplating lingering over the intermediate periods for each of these cases."[41] Thus, Gallimard now agreed to include in the series work on the world outside Europe but still implied a somewhat more systematic series than that which Bloch had in mind.

For his part, Bloch envisioned a series in which the books focussed on particular themes or "problems" in a great variety of settings – ranging from very local ones such as the Limousin (which Gallimard felt was too narrow) to his own treatment of both France and England. As Bloch wrote to Febvre following the first meeting, "I would not want stereotypic volumes, the German – English – Saint marinois peasant . . . etc." He speculated, for example, on a book by Georges Lefebvre on "the French peasant and the Revolution," one by Kan'ichi Asakawa on "the Japanese peasant," one by Gabriel Le Bras on "the church and the village," one by Febvre on the Jura, one by André Siegfried on "the political life of the peasant of the 'Midi'" and an unspecified book on Germany of which he remarked: "But I still do not see very well the theme to be treated (East and West – Revolts and Reformation?)."[42] Bloch agreed with his publisher on the need to "reinforce a little the portion on France" but noted that it would not be easy to find the appropriate authors[43] – an indication, perhaps, that he did not intend to draw heavily on the Vidalian regional geographers, who had concentrated on regional studies of France leaving the studies of other places to ethnographers and others.[44]

In a related discussion, Bloch objected to Gallimard's title, "History of the Peasantry": "I do not particularly like either 'Peasantry,' or even 'History.' My preferences would be something like: 'Le paysan et les champs'."[45] To Bloch, the use of "History" may have implied both a more comprehensive series than that which he planned and an approach which did not necessarily draw on the other relevant "human studies." "Peasantry" perhaps appeared too abstract and rigid. By contrast his title "Le paysan et les champs" (the peasant and the fields) reflected his interest in the "concrete living reality." The title finally adopted, "Le Paysan et la Terre" (the peasant and the earth) shifted attention from the social landscape to more geographical themes and seems a weaker reflection of Bloch's approach. Unfortunately, it is not clear how and why the final title was adopted. Bloch in any case stuck to "Le paysan et les champs" during 1935 and the new title does not appear in his file of Gallimard correspondence until the war years when Bloch, living in the free zone, had a diminished role in the direction of the series.

Despite Bloch's reservations over the use of the word "history" in the title for his series, he drew very largely on social historians, often those with a serious interest in Durkheimian sociology. Some additional help was obtained from ethnographers and geographers who also had an interest in social phenomena and Durkheimian thought. By contrast, during the war years when Bloch had less control, geographers with no particular interest in Durkheim predominated among the new candidates.[46]

Bloch's preference for a problem-oriented social history was evident in his correspondance with potential contributors. For example, Jean-Paul Hütter wrote in response to a letter from Bloch, "You tell me that you are not looking for the 'Grundriss' that wants to say all, but for the book which poses the principal problems."[47] Sion, in turn, wrote:

After having read your letter, I see even more clearly that it is not a question of volumes in a series of the same type, but of works each capable both of standing on its own, organized according to its own laws, and if possible combining with others "in order to throw *feux croisés*" [crossing lights or cross fire], as you put it so well.[48]

In addition, contributors were permitted to restrict their study to a given period so long as it was adequate to explore their problems of interest.[49]

Despite his advocacy of a flexible approach, Bloch encouraged the adoption of a comparative framework where possible. For example, when writing to Henri Labouret, Bloch suggested that Sudan might form too limited a topic, and for Scandinavia he envisioned a book on "the peasant of the North" rather than one just on Denmark or Norway. He hoped, he explained, that such a book would address the question of the lack of seigneuries there, a question of general interest to European historians. Although Sion initially wanted to restrict his book to the French Mediterranean, after some prodding he finally agreed that Bloch was justified in advocating a larger area allowing more room for comparison with other Mediterranean places.[50]

Just as Bloch discouraged a limited regional framework, he also wanted a focus on man not agriculture *per se*.[51] The point of view, however, was not to be focussed on any man but on the peasant. As in *La Société féodale*, Bloch advocated a study which highlighted the key characteristics of rural society. In a letter to Lis Jacobsen, he explained, "'History of civilization.' Yes, most certainly. In addition to the social structure and the agrarian life in the strict sense, the *mentalité* will have a large place – the accent always put, of course, on the rural aspect of society."[52]

Bloch also encouraged some authors to explore religious phenomena. He explained to Jacobsen that he did not see the series as simply an economic one and hoped that, like Le Bras, she would pay particular attention to religion. On reviewing Labouret's outline, however, Bloch objected to his designation of a separate chapter on religion, arguing that religious phenomena, which were "closely linked to all the manifestations of social life," could be treated more effectively along with the other social phenomena to which they were tied.[53]

The series was introduced in 1941 with an unsigned "Note from the editor" in Labouret's volume. There Bloch spelled out the aims of the series, returning to his metaphor of *feux croisés:*

It is to the description, analysis, and explanation of the various types of peasant humanity, that this series of volumes is devoted . . . In a word, by the flexibility of the overall design and the diversity of the works which we conceive as so many *feux entrecroisés* [intersecting lights], we have done our best to do justice to the very variety of life.[54]

In this introduction, Bloch adopted an approach that was closer than ever to Durkheimian sociology; he not only wrote of types but also drew on the concept of "civilization." Nevertheless, he kept his interest in diversity and variety over laws and generalizations and implied an approach which studied a given society within its historical context.

Elaborating on his metaphor of *feux croisés,* Bloch began to articulate the moral vision that supported it. He expressed his belief in "the attraction and sacred character of any human effort under whatever skies or by whichever branch of the great family of men that one accomplishes it." Although the series was not to depict "peasant realities" in the "faded colors of an idyll," Bloch suggested that human life had proved more powerful than the events which challenged it.[55] The message here was one of hope and resilience during what were exceedingly disrupted times.

During Bloch's lifetime only one of the several volumes he had planned appeared, Labouret's *Paysans d'Afrique Occidentale*.[56] Despite its ethnographic character, Labouret's book examined many of the same questions as Bloch's *Les Caractères originaux*, as it centered on social life and the village and even made explicit comparisons between medieval Europe and contemporary Africa. Bloch appears to have approved of Labouret's wide-ranging comparisons, though he would later object to some of Labouret's comments relating to the European nobility.[57] On receiving his reactions to the manuscript, Labouret wrote to Bloch, "And since you did not react too strongly when I carefully put forward a few remarks on the customs of Western Africa and those of the European Middle Ages, I will, no doubt, pursue the matter in connection with the collective lands and individual rights."[58] Paralleling Bloch's approach in the *La Société féodale* and elsewhere, Labouret talked of collective psychology, noted the complicated land ownership rites, and contrasted oral and written traditions of law. Furthermore, in discussions of the "habitat," he was not content simply to provide a description but instead tried to relate the habitat to characteristics of the *genres de vie*, the family, and the social organization.

For Bloch's own contribution to this series, he had proposed a work entitled "Seigneurie française et manoir anglais." This work was never completed, but Bloch did give a course at the Sorbonne and a series of talks at the London School of Economics with the same title and also undertook a comparative "seigniorial inquiry" in the *Annales*. Bloch's course began with an examination of the contrasting contemporary rural landscapes of England and France. He insisted that those contrasts "were not inserted in nature" and turned instead to economic and social factors. Describing the seigneurie as "an economic enterprise," he argued that similar economic conditions in the two countries had had differing effects "because they act on very different social substrata."[59] Bloch's use of the term "social substratum" here was broader than the Durkheimian usage since he implied that the key element was a difference in legal structure; in Durkheimian terms that structure would have been considered as an institution, not part of the "social substratum."

When Bloch was not so tightly constrained by his role as a professor of economic history, he phrased the problem of the comparative history of the seigneurie in more clearly social terms. The key question, he claimed, was "the variable relationships of the seigneurie and the community." To address it, one must look at the seigneurie not only as an economic enterprise but also as "a group based on command," examining "the superimposition of the power of a man on the ties of the community." Furthermore, he argued that European history by itself could not give one a sufficient understanding of the seigneurie, in part due to the lack of documentation. Noting the role of the comparative method in eliminating certain explanations (as Langlois and Seignobos had done), Bloch also saw a more positive (albeit modest) role – that of suggesting new interpretations. Nevertheless, he continued to argue that its main value was to uncover differences.[60]

In his direction of the series "Le Paysan et la Terre" and his related writings, Bloch demonstrated an even wider perspective than that shown in his own substantive work, choosing examples from around the world and from very different historical periods, a perspective that was also apparent in his speculations on future projects. Advocating a flexible approach in order to address the particularities of the situation studied, he proposed to highlight differences. He retained his interest in interdisciplinary problem-centered studies and encouraged a comparative study of social life with particular attention to religious phenomena. Thus, strictly regional frameworks were discouraged as were studies focussing on agriculture let alone the land. The monographs were not to scan the entire world and all of history but to make selected comparisons – whether that be between neighboring and contemporary societies, as in his own plans for a study of the French seigneurie and the English manor, or between contemporary Africa and medieval Europe. He hoped with his *feux entrecroisés* to give a deeper understanding of "the very variety of life."

In addition to his work for Berr and his own series, Bloch had several other projects in mind that also illustrated his broadening perspective. In October 1939, he wrote to Febvre of his plans for a "History of French society within the framework of European civilization." He began a first chapter entitled "Naissance de la France et de l'Europe," and wrote a methodological introduction in which he reiterated the principles of documentary criticism. According to Bloch, applied psychology and probability theory had added to the basic ideas set down by seventeenth-century scholars. However, one should keep in mind that the human spirit was constantly changing, that the only universal characteristic of the human mind important for historiography was its inability to remember clearly, and that psychological and social factors limited one's ability to rely on mathematics. Characteristically, he also noted the complexity of the issues at hand – in this case confused by the mixing together of the notions of state, nation, and civilization. As the proposed title indicated, he intended to address a very social question, but as before his approach was to be an historical one. Although the work was to be a history of the French as a social group examining both popular and ruling classes, he would exclude neither political history nor individuals. Such a study, he argued, must place French society in the context of European civilization because this is the only way that "it would be possible to measure a national originality, for which the exact image could only come from the perception of

contrasts." More than ever, he seemed convinced that one could only understand France by placing it in a much broader context.[61]

Later letters alluded to a possible "total rewrite of the *Caractères*" and a
history of the First Reich.[62] He even speculated on the interest of a work on
the United States:

> If I were young, once having put matters in order on this side of the great pond (pious
> wish!) I would go to the other shore for three years in order to gather the materials for
> a history of *peuplement* of the United States. What a passionate subject! and one
> which, I believe, only an historian familiar with the rural realities of our old earth
> would be capable of seeing almost through . . . [63]

He now seemed ready to leave his French base and tackle questions relating to
other places shaped at least in part by the same Western civilization.

The last work that Bloch began – what has become known as *Apologie pour
l'histoire ou métier d'historien* – was a general work on the nature of history,
or as he described it, "an artisan's notebook" rather than a philosophical
work.[64] This book, begun by the summer of 1941, took up themes that had
intrigued Bloch for some time.[65] Febvre was not very enthusiastic about
Bloch's plans and seemed worried that his book would resemble the classic
French work on historical methodology by Langlois and Seignobos. Bloch, in
turn, replied, "Poor father Seignobos! poor Charles V. Langlois . . . There are
moments when I ask myself if these satans do not haunt your dreams. Don't
worry: if 'métier d'historien' is ever written, it will not be under their sign."[66]

The unfinished manuscript left by Bloch did indeed differ significantly from
Introduction aux études historiques by Langlois and Seignobos and also
demonstrated how far his position had progressed since his 1914 lycée address
at the Lycée d'Amiens. Although some of these differences reflected Simiand's
1903 criticisms of history, Bloch's book was clearly not written under a
Durkheimian "sign" either. As Bloch wrote to Febvre in 1943:

> I have continued to write my "Apologie pour l'histoire." A difficult exercise. An instruc-
> tive exercise. I still do not have a very precise idea as to what the book will be, or even
> if it will ever be. But it is never a bad idea to plot one's position. Durkheim was cer-
> tainly not an imbecile. Neither was (hide your face!) the poor father Seignobos. Nor
> Charles V. Nevertheless, how far we are from them. If it were only in our solutions or
> our attempts at solution, it would still be nothing. But it is even in our problems.[67]

And indeed, in contrast to his 1914 address, Bloch did go much farther in
articulating what separated his approach from those of Langlois and
Seignobos and of Durkheim.

In his manuscript, Bloch suggested that both Seignobos and Durkheim had
been marked by the Comtian view of science which dominated the late nineteenth and early twentieth centuries. He claimed that when applied to the "historical studies" such a view had led to two tendencies. The first was that of *les
historiens historisants* who were marked by "disillusioned humility."

According to Bloch, they could be characterized by the unfortunate statement of Seignobos, "It is very useful to ask oneself questions, but very dangerous to answer them."[68] Bloch, by contrast, tried to give some answers in his problem-oriented studies even if conjectural ones and stressed the need to articulate the limits of one's understanding.

The second tendency was represented by Durkheimian sociology. According to Bloch, the sociological school founded by Durkheim "believed it possible, in fact, to institute a science of human evolution which was modelled on a sort of pan-scientific ideal, and they did their best to establish it."[69] The price that was paid was to dismiss under the rubric of "event" many "very human realities" and a "good part of the most intimately individual life" – although Bloch did admit that attempts had been made to soften the "initial rigidity." Despite his reservations over some of their positions, he felt that much was owed to the Durkheimian school. As he explained, "It taught us to analyze more profoundly, to grasp problems more firmly, and to think, if I dare say it, less cheaply. It will only be spoken of here with infinite gratitude and respect. If it seems superseded today, that is the price that all intellectual movements must pay, sooner or later, for their fruitfulness."[70]

To clarify the difference between his approach and that of his predecessors, Bloch drew analogies to both Albert Einstein and quantum mechanics. Through their influence, he argued, the notion of science had changed relying now on concepts of probability and relativity rather than rigorous measurement. Furthermore, he rejected a uniform model for science based on the physical sciences, arguing, "We do not yet know very well what the sciences of man will be one day. We know that in order to exist – while continuing, it goes without saying, to obey the fundamental rules of reason – they will not need to renounce their originality, or to be ashamed of it."[71] Human phenomena, he argued, were very delicate and as such often impossible to measure. To "translate" and by so doing to understand them, one must often "feel with words." In contrast to the Durkheimians, Bloch made it clear that he was not particularly interested in laws of repetition. Instead, his concern was that of differences and relativity. The most the historian could hope for, he claimed, was the establishment of multiple "causal wave trains," which in turn would be limited by one's particular perspective of inquiry.[72] Even this, however, was more than *les historiens historisants* sometimes attempted. Durkheim, by contrast, had aimed for a more ambitious form of explanation than that attempted by Bloch and did not share Bloch's fascination with and celebration of the "different."

Bloch's form of history also differed from both that promoted by Langlois and Seignobos and from Durkheimian sociology in its object of study. Langlois and Seignobos had stressed prominent persons and events and had suggested that the social could and should be reduced to individual acts. Significantly when speaking of the social, Seignobos usually spoke of social action – the social was seen as a cause of particular events rather than as a psychological phenomenon.[73] Bloch, on the other hand, criticized a history focussing on events and promoted a more social history. In 1935, Bloch had, in fact, noted that the difference between his approach and that of Langlois and Seignobos was largely one of the phenomena studied – this in response to Febvre's charge that Bloch's use of the term *juge d'instruction* smacked of *Seignobosisme*. Then, he wrote to Febvre, "If my metaphor repels you, that is because you immediately imagine an individual trial. . . . But the confrontation of witnesses, cross-examination, etc., etc., does not necessarily mean a debate over persons or minor events."[74]

Despite significant parallels, Bloch's treatment of the social was never strictly Durkheimian. In *Apologie*, he returned to the example of the Revolution of 1848 disputed by Seignobos and Simiand and appeared to side with Simiand. In 1914, he had used the same example to illustrate how to deal with contradictory testimonies. This time Bloch argued that this example did not represent "the very type of the fortuitous event"; instead he stressed that the shooting on the Boulevard des Capucines represented only the spark which ignited what was already a very explosive situation. By implication it was the underlying social causes rather than the events which were significant. Also, just as Simiand had insisted in 1907 on the importance of "objective psychology," Bloch indicated that to him social conditions were "in their profound nature, mental" – a contrast in both cases to Seignobos' portrayal of the social as the cause of particular events.[75] Nevertheless, Bloch remained unwilling to separate the social and the individual. Charging that the Durkheimian sociologists viewed history as simply a depository for "the most superficial and fortuitous human phenomena," he argued that he would adopt its broader meaning, ruling out neither the individual nor the society, neither momentary crises nor "the pursuit of the most durable elements."[76] Unlike Durkheimian sociology, history as conceived by Bloch would study both the social and the individual, leading to a mixing between these two that would have been unacceptable to Durkheim.

Related to Bloch's unwillingness to separate the social and the individual was his position on teleology ("finalism") and "agents." In contrast to Durkheimian sociology, Bloch continued to argue that "the science of men" was justified in studying the motives, aims, and goals of conscious human beings. Although physical science excluded such words as "success" and "failure," "awkwardness" and "skillfulness," they were appropriate for the study of history. Similarly, Bloch did not totally dismiss the concept of human agent and at times when discussing human action, appeared to endorse this concept. Unfortunately, however, a clear discussion of this matter was not given in *Apologie*; it seems likely that it was to be included in a later section on explanation in history.[77]

Apologie also differed from both the early manual of Langlois and Seignobos and from Durkheimian sociology in the type of concepts which it promoted. Bloch paid particular attention to the classification of historical time. Not surprisingly, he did not defend the periodizations used by historians of his father's generation, which Simiand had so strongly criticized in his attack on the "chronological idol." Instead, Bloch wrote of "the disarray of our chronological classifications" singling out for attention divisions by reigns and governments and attacking them for their "false exactitude."[78]

Bloch's alternatives reflected not only his interest in social phenomena, but also his view of science. To designate successive stages for "social evolution," Bloch suggested the use of two "flexible" concepts: generation and civilization. Of the former, he noted an irregular cadence which matched that of "the social movement." He also liked the possibility of an interpenetration of generations which reflected both differing influences on different groups and the effects of "personal temperaments." Seeing this lack of simple definition as a strength not a weakness, Bloch wrote: "The notion of generation is thus very supple, like every concept which endeavors to express, without distortion, human things. But it also responds to realities which we feel to be very concrete."[79] Similarly, the concept of "civilization," which could be used for longer phases, reflected "tonalities" which were inherently difficult to express:

In sum, human time will never conform to implacable uniformity such as the rigid division of time by the clock. It needs measures which are attuned to the variability of its rhythm and which, in accordance with reality, must often rest content with the identification of marginal zones for boundaries. It is only by this plasticity that history can hope to adapt its classification to the "very lines of the real," to use Bergson's words, which is properly the ultimate end of any science.[80]

Bloch now viewed science in a rather different light from his predecessors. Even though he would not follow the generation of 1905 in totally rejecting the principles of the Nouvelle Sorbonne, he was apparently impressed by some of Bergson's teachings.

When discussing more specific concepts such as "feudalism" and "capitalism," Bloch cautioned, nevertheless, against stretching their meaning to encompass what he saw as fundamentally different realities. Flexibility was not meant to give a license for sloppy analysis and an abuse of terms. Unfortunately, according to Bloch, both "capitalism" and "feudalism" had been weakened in this way.[81] The fault, he explained, lay in the historian's reluctance to define his concepts. Bloch would speak of stages of evolution but remained very suspicious of the general labels associated with them. Despite his association with the sociologists, he remained interested in "originalities." Accordingly, for Bloch, many concepts would retain their base within a particular historical context, something which, he felt, even historians could be faulted for overlooking.

Nevertheless, Bloch did not urge a simple adoption of the terminology associated with a particular time and place. Such vocabulary in his view should be viewed as "testimony," subject to criticism and analysis using the principles of historical semantics. Despite abuses of terminology, he argued that the search for common labels was essential. On the other hand, given his sensitivity to differences and originalities, Bloch leaned toward a heavy use of the terminology of the time and place. As he explained, "Certainly, despite everything, the names, however incomplete their adherence, hold much too strong a grip on the realities to ever permit a description of a society without a considerable use of its words, duly explained and interpreted."[82]

In brief, Bloch strove for sometimes conflicting ends. He wanted flexibility in his concepts and at the same time both a lack of ambiguity and adherence to concrete realities. He wrote, for example: "For at the outset, any analysis wants an appropriate language as a tool: a language capable of drawing the contours of phenomena with precision, while conserving the necessary flexibility to adapt itself progressively to discoveries – a language without vagueness and ambiguities."[83] Bloch's search for a happy medium between *les historiens historisants* and the Durkheimians was bound to be a difficult one and was fraught at times with contradictory tendencies which he did not always articulate.

Another aspect of this dilemma was evident in Bloch's attraction to both the abstract categories of sociology and the concrete realities of history. Whereas Seignobos had attacked "pure abstractions" particularly those with a social content, Bloch argued that abstractions were central to the historian's method.[84] He was not, however, willing to abandon his interest in "concrete realities," an interest which he attributed, in part, to the teachings of Seignobos. For example, in 1938, Bloch wrote to Febvre regarding what he saw as the latter's overly severe criticisms of Seignobos:

I have never hidden from you that I hold, having been his student (which was never your case), some gratitude and some affection for this old man who has always been kindly toward me; who, having never counted me in his little court at the time of his power, had the good taste, nevertheless, to show his friendship toward me on several occasions; above all, who amongst my teachers at the Ecole and the Sorbonne was, my father aside, the only one to strive to interest us in a history not completely empty of men and realities; and finally, who recently reminded me of these memories in a little note whose tone moved me.[85]

Bloch's "concrete realities" were not however identical to those of Seignobos, being more intimately associated with people's thoughts. In this way, at least, these realities were somewhat closer to the concerns of the sociologists. In contrast to Langlois and Seignobos, he showed a greater tolerance for the "abstract" concepts of sociology and for periods marked by fluid boundaries, and in contrast to the sociologists, he was more willing to use and promote historically based concepts, rather than those which could be used for all places and times.[86]

Specific reflections in *Apologie* on geography as a discipline were rather limited. The main discussion came as an illustration of the limitations which a given perspective imposed on one's understanding:

"Anthropogeographie" studies societies in their relationships with the physical milieu: mutual exchanges, obviously, in which man continuously acts on things at the same time as they act on him. In this case again, one has nothing more or less than a perspective whose legitimacy is demonstrated by its fruitfulness, but which other perspectives will have to complete. That is, indeed, the role of analysis in any category of research. Science only decomposes the "real" in order to observe it better, thanks to an interplay of "feux croisés" [crossing lights] whose rays constantly combine and interpenetrate. The danger begins only when each projection pretends to see all by itself; when each canton of knowledge takes itself for a "patrie."[87]

Thus, Bloch did not deny the legitimacy of *anthropogeographie* as a field of study and yet used it as an example of the dangers of relying on only one point of view. It is interesting that he used what was by then the old-fashioned term *anthropogeographie* to make his point. A later reference to "the pseudo-geographical determinism, today definitely ruined" also illustrated both Bloch's suspicion of simplistic geographical explanations and his recognition that his implicit criticisms were less true of *la géographie humaine.* Speaking of complex interactions between people and their environment, he elaborated, "In nature, is not man the greatest variable?"[88]

Earlier in the manuscript, Bloch argued that in order to understand the complexities of such an apparently geographical topic as *l'occupation du sol,* one must examine a great variety of evidence including associations of vegetation, archeological finds, medieval charters, and linguistic evidence found in place names. Bloch continued, "Few sciences are, I believe, forced to use so many dissimilar tools simultaneously. This is because human phenomena are particularly complex; because man finds his place at the summit of nature."[89] Analysis of each of these forms of evidence required its own particular skills and techniques, making the researcher's task inherently difficult and so would often require a close collaboration between scholars.

As indicated in both of these references, Bloch's real interest lay in people. "The greatest variable" in nature was man and a study of the occupation of the soil was seen as one of "human phenomena." The evidence of geography, as in an examination of landscape, was to him merely one way, albeit an important one, of reaching the underlying human realities. Returning to the blood and flesh metaphor, he wrote:

Behind the perceptible features of the landscape, the tools or the machines, behind what appear to be the stiffest writings and the institutions most completely detached from their founders, it is men that history wants to seize. Whosoever fails to reach them will never be more than a laborer for erudition. The good historian resembles the ogre of the legend. Wherever he senses human flesh, he knows that there lies his prey.[90]

As in his earlier writings, Bloch in *Apologie* saw the landscape as a symptom of something else, but believed that it could nevertheless provide a useful point of departure. When discussing the links between the past and the present, he suggested that a landscape such as that of the open fields of the north of France could only be understood by a careful examination of social history. On the other hand, to understand that early history an examination of the present landscape was required – not to give all the answers but at least to provide a useful starting point.[91]

Despite the value which Bloch placed in a study of landscape, he remained somewhat uncomfortable with the notion of landscape as an object of study. Following his discussion of the angle of sight of human geography and the danger of an overly narrow view, one finds a somewhat confusing passage in which Bloch argued that the unity of a landscape lay only in the mind of the observer. Bloch continued that whereas the natural scientist tried to abandon the observer in order to better understand the observed, the object of history, and by implication "the human sciences," was "human consciousnesses." In 1934, he had argued that there was no unity of place – that instead the focus should be on a problem. Bloch now seemed to imply that the landscape might possibly form a subject but only if one understood by "landscape" a perceived landscape – that is to say a manifestation of "human consciousnesses."[92]

At first glance, Bloch's approach in *Apologie* seems very much like that of his father Gustave, as expressed in the debate at the Société Française de Philosophie in 1908. Like his father, he rejected both the skepticism and timidity of Seignobos and the use of laws of repetition in history advocated by the sociologists. However, in contrast to his father, he saw a greater resemblance between history and the other "human sciences" even though he did agree that history was fundamentally different from the natural sciences. Also, he would not have described "the phenomena of human geography" as "very simple and very crude" and as such susceptible to formulation in laws. For him, in contrast to Gustave Bloch, the phenomena studied by human geography were those which his father had termed "the psychological phenomena" that were "so diverse and so complex."

In this work, Bloch illustrated more clearly than before how he differed from his teachers. In contrast to "father Seignobos" and "Charles V.," Bloch searched for explanations, did not fear abstraction, saw the most interesting history as social history, and had a greater tolerance for periods marked with fluid boundaries. Nevertheless, he kept an interest in differences and the concrete, even while transforming the meaning and implications of those words to make them both more social and more tied to human consciousness. Although he did not totally reject the ideal of explanatory science, he had a rather different view from the Durkheimian sociologists. He saw science as far more relativistic and flexible and valued the highlighting of differences. Bloch's form of history also contrasted with Durkheimian sociology because of Bloch's unwillingness to separate the individual from the social, his belief that history should study the motives and goals of conscious human beings, and his promotion of historically based concepts. Although he could not fairly be labelled *un historien historisant*, he was clearly not a Durkheimian either.

Among the problems which continued to intrigue Bloch during these final years were those suggested but not defined by geography. Despite all his work on money, he retained an interest in rural history – contemplating a thorough rewriting of *Les Caractères originaux* and even speculating on the interest of writing a history of the settlement of the United States. To Bloch, these were problems of social history in which geographical evidence helped to give some life and concreteness. From Durkheimian sociology he had taken a fascination with social phenomena, which became incorporated into his object of study, but he never approached that object in a truly sociological way. From Vidalian geography, he developed his interest in landscapes and the "concrete" – but only to understand what he saw as the underlying social reality. Geography was to him just one of many sources for his *feux croisés*.

10

Towards a reworking of the historiography of Marc Bloch

Even in his later years when he came closest to Durkheimian sociology, Marc Bloch remained essentially an historian. He was an historian in the sense that his primary interests lay in change and differences rather than laws and theory and that the problems which he chose to address were human ones rather than those of the physical environment. Although not entirely immune from the scientistic atmosphere of the Nouvelle Sorbonne, his vision of history as science was that of the critical documentary method which then defined professional history, rather than the law-seeking of the sociologists or the lapse into biological metaphors of the geographers.

Nevertheless, an understanding of Bloch's relationships to the fields of sociology and geography is essential for an understanding of his approach; although that approach remained historical, the form of history which Bloch developed was deeply marked by those relationships. By transforming history to address some of the concerns brought forward by the sociologists and geographers, Bloch was able to diffuse some of the criticisms his discipline faced. To date, these issues have not been central in the historiography of Marc Bloch, weakening both a comprehension of the particular character of his approach and of how his approach came to be. Both by focussing on Bloch's relationships to these two fields and by paying greater attention to the institutional contexts within which he wrote, the understanding of his approach can be enriched.

In general, the historians who have written on Bloch have been primarily concerned with exploring Bloch's work in relationship to other historians, rather than with those from different fields. Among those discussed have been members of the French historical establishment which Bloch helped to challenge, others associated with the "Annales school," or foreign historians, especially German, English, and Belgian. Many have also been taken with Bloch's active participation in both world wars, in the Resistance, and his subsequent martyrdom. Discussions of his method have stressed Bloch's breadth – his wide reading in many languages, in many fields, and from many periods, and his marshalling of previously untapped or underused sources such as the *plans parcellaires*, the forms of plows, and relics in the landscape and in language. Accordingly he has been depicted as a forerunner of the so-called *histoire totale*. His use of provisional synthesis without overly artificial classifications, his problem-orientation without judgment, and his use of comparisons have also been highlighted.[1] On the whole the consistency of Bloch's approach rather than its development have been stressed.[2]

For the many sociologists and geographers who have also admired Bloch's work, his relationships to their disciplines have been more central, but, for the most part, they have done little to examine his work directly, apparently leaving that task to the historians. Instead, they have often used Bloch's works in teaching, and used his approach to justify their own. In recent years, the greatest attention from sociologists has come from those attempting to establish the basis for historical sociology. For example, Theda Skocpol has classified Bloch as an "analytic" historical sociologist, to exemplify the type of approach which she prefers. According to Skocpol, such an approach seeks "causal generalizations" rather than the "complex particularities of each place and time" associated with interpretative sociologists or the "general models" associated with the last of her three groups. Similarly, Daniel Chirot labelled Bloch as "one of the founders of contemporary historical sociology" and implied that he favored "broad comparative generalization" and even "sweeping theoretical conclusions" over a "pedantic attention to detail." Denis Smith has also put Bloch among the ranks of the central "historical sociologists." However, in contrast to Chirot and Skocpol, Smith was more attentive to the historical character of Bloch's approach; he classified Bloch's approach as falling midway between one favoring the "primary exploration of specific historical situations" and one of "empirical generalization" but not as one of "systematic theorizing."[3]

References to Bloch's work abound in the works of geographers in France, Britain, and North America. In three broadly ranging essays, Hugh Clout, Alan R. H. Baker, and Paul Claval explored the relations between geography and the Annales school of history, but, in the interests of generalization, they paid little attention to differences within the Annales school. Accordingly, the significant criticisms which Bloch, in particular, made of geography were not highlighted.[4] In a number of other cases, Bloch's work has been singled out by geographers who like the historians have been impressed with his interdisciplinary approach.[5] Due to differing histories of disciplinary development, the use of Bloch's work by French geographers has differed from that by geographers in Britain and North America.

In France, given the triumph of Vidalian human geography over the earlier *géographie historique*, Bloch's work has been more often used in teaching and discussed in relationship to debates over the agrarian structures than cited in general statements on the geographical approach as an example to follow.[6]

The historian Georges Duby, for example, has testified that he was first introduced to Bloch in the late 1930s by his *maîtres géographes,* but it is to Roger
Dion, often with little if any mention of Bloch, that the French geographers
have turned in an attempt to define and promote historical geography.[7]

In addition to contributing to the study of Bloch's role in the French debate
over agrarian structures,[8] British and North American historical geographers
have often drawn on Bloch's work in more general statements on their method
in an attempt to orient their relatively young discipline, which dates only from
the late 1920s to early 1930s.[9] This was particularly the case in the 1970s and
early 1980s as they reacted to the quantitative methods and model building
that then dominated parts of Anglo-American geography. His name has been
invoked in support of very different approaches leaving one with a confusing
and even contradictory message. Examples range from those relying on the use
of model-based paradigms to those of synthesis, the understanding of the
particular, and a poetic interpretation of regions and places. At times, his work
has also been used to promote various techniques such as a retrogressive
approach and an examination of intervening periods and at others to encourage a shift in the object of study giving more place to human consciousness,
tradition and social struggles.[10]

Although the questions of Bloch's relationships to sociology and geography
have not been central to historiographical work, they have not been entirely
ignored. Many authors have both noted his debts to these fields and remarked
that of the two, sociology was the most influential.[11] However, the image
created is a mixed, cursory and even misleading one in part due to mis-
apprehensions about the development of both geography and sociology and
a related tendency to ignore the institutional context within which Bloch
worked. For example, the important distinction between historical and human
geography in France has been blurred, leading some to exaggerate the historical and "humanistic" character of the latter and accordingly both to mis-
represent Bloch's relationships to Vidalian geography and to ignore his
assessments of French historical geography.[12] Bloch's interest in economic as
opposed to social history has been exaggerated as some have failed to see the
links with his interests in sociology or the institutional constraints which
increasingly shaped Bloch's writing.[13] The marginal character of Bloch as provincial rebel has been overstated as has the novelty of his approach.[14]
Furthermore, the roles of generalization, explanation, and theory have been
exaggerated as later authors have attempted to use his work as a precedent for
their own approaches.[15]

The increasingly accepted interpretation that Bloch was closer to
Durkheimian sociology than Vidalian geography has merit, but still needs
considerable qualification. The similarities between his approach and that of
Durkheim have often been exaggerated. For example, it has been claimed or
implied that Bloch's comparative method and use of certain key concepts were

Durkheimian when in fact they were clearly not.[16] Commentators have frequently overlooked how Bloch's mixing of the social and the individual went directly against Durkheimian principles. In addition, Bloch's effort to give an "internal" view of historically based social groups and his corresponding criticism that some of the work by the Durkheimians was monolithic has at times been ignored.[17] Furthermore, little has been done to trace Bloch's changing relationships to Durkheimian thought, though a few observers have hinted at it.[18]

Discussions of Bloch and geography have typically been one-sided either stressing perceived parallels between Bloch's approach and that of the Vidalians[19] or noting his criticisms of the Vidalians with no clear discussion of what it was that attracted him to the field.[20] Observers have also overlooked how Bloch's use of geographical evidence differed from that of the geographers.[21] Finally, though occasionally intimated, very little has been written on Bloch's attempt to portray the "concrete living reality," a goal which helps both to distinguish his work from the Durkheimians and to demonstrate what it was that he found useful in the works of the Vidalians.[22]

Here Bloch's work has been examined in the context of the disciplinary debates and change involving two fields, sociology and geography, that were particularly important in shaping his approach to history. Very much remains to be done to complete a "direct" historiographical study of Bloch as has been recommended by Massimo Mastrogregori or to examine the context in which Bloch's thought evolved, as Olivier Zunz has advocated. A more thorough examination of Bloch's relationships to history and to the many other fields on which he drew is still needed. In addition, much remains to be done on the links between Bloch's work and the broader social, economic, and political contexts of his time.[23] Nevertheless, an examination of how Bloch met the challenges posed by the changing fields of sociology and geography, can shed light not only on his objects of study but also on his terminology and the methodological questions which he addressed. This holds not only for *La Société féodale*, which some historians such as Bryce Lyon have criticized as being too sociological, but also for his other works including *Les Caractères originaux*, which Lyon championed as his best work.[24]

The ways in which Bloch dealt with the challenges which geography and sociology presented enabled him to develop an historical approach that was neither Vidalian nor Durkheimian and yet set him apart from other historians as well. Intrigued by the teachings of Vidal and Gallois during his days at the ENS, Bloch held them in respect for the rest of his life. Nevertheless, following Simiand's attack on the field, Bloch came to reject much of their teachings. He initially proposed a regional framework for his thesis, but as early as his monograph on the Ile-de-France, he rejected the idea of region as an object of study or a real entity. Instead, he viewed the study of region as just a means to a different end, and a rather limited approach at that.[25] A regional approach was only appropriate in certain cases and even then needed to be socially and historically defined and, as he increasingly argued, to be supplemented by comparative work.

Attracted to a study of landscapes and using them as a point of departure for some of his own work, Bloch's object of study was to be that of man in society and never, as Vidal had proposed, the terrestial physiognomy. Because of his interest in the human character of landscapes, Bloch was drawn to the study of "habitat" and field systems. However, to understand such topics, he turned to sociology, social history, and historical geography, but never, as Vidal had done, to plant ecology. Rather than examining landscapes to uncover an explanatory *enchaînement* leading to geographical laws, he hoped, as he eventually explained, to add some concreteness to his understanding of a particular social reality. He even went so far as to refuse to allow geographical factors to enter into his explanations of field systems. Also rejecting Vidal's suggestion that geography was the study of places, Bloch argued that work should be devoted instead to the study of particular problems – problems which Bloch saw in social terms and which, he gradually admitted, geographers were often poorly placed to answer.

Of all the Vidalian concepts, Bloch was most attracted to that of *genre de vie*. He praised it in Febvre's early work on Franche-Comté and in Halbwachs' later work on suicide, and he used it in his own work on social class and feudal society. As was so often the case for concepts that Bloch found attractive, this term was never clearly defined, either by him or in this case by Vidal. Despite its Vidalian origins, one suspects that it had, for him, a largely social content. For example, in his article on the nobility, he used the term to describe not only the habitat, armaments, and occupations as Vidal might have done, but also the alliances, patrimonies, and even more generally, the "culture" and "way of thinking." Perhaps to Bloch, the lack of precise definition gave the concept of *genre de vie* the *souplesse* needed to describe complex social realities.

For Bloch, the most important contribution which geography could make was to help give a concreteness to the study of what he saw as a fundamentally social reality. This was not, however, in his eyes a minor achievement. With his fascination with "originality" and difference, such a concreteness was deemed essential to the task at hand. Only by using such detail, he argued, could one be "exact" and close to life. In his own substantive work, he drew heavily on what could be seen as geographical detail – but in this case such detail was carefully selected for its "human" and "social" significance. Regional names and identifications chosen were those used by the people of the times. Landscape description stressed the habitat, terrain, climate and, of course, the *genres de vie*. A study of landscape and *plans parcellaires* could give one clues not only to important issues of social history, providing in this way a useful point of departure, but also helped at the end of a study as "the most immediately perceptible image of profound social realities." For a particular place, one could investigate how the townscape and town plans translated its history, a history which to Bloch was inherently social. Although he championed concreteness and direct observation, the detail which he used was carefully selected and never used as an end in itself.

Identifying somewhat more closely with the field of sociology, Bloch took longer to separate himself from it. He developed an interest in rites, religious phenomena, and serfdom as a social institution while at the Fondation Thiers, where he worked closely with Davy, Gernet, and Granet. However, due to the interruption of the First World War and no doubt also due to the constraints of his profession, he was unable to concentrate on these interests in his thesis. Following his thesis defense, he turned directly to the study of a rite, the royal touch, only to receive fairly harsh criticism from Halbwachs for his chronological framework and narrow topic. Returning to some of his dissertation research on the freeing of serfs, Bloch attempted in a series of articles both to examine serfdom as a changing social institution and to investigate its class character. Like Halbwachs and Simiand, he approached social class through collective representations even though he remained somewhat more cautious than Halbwachs in applying the label of class to medieval groups. Class remained for him just one of many important social classifications. These changing classifications were, he insisted, exceedingly difficult to pin down and could only be understood through a careful examination of the language created and used by the people of the times.

Bloch turned more directly to methodological questions in his writings on the comparative method and his reviews of the Durkheimians, and there he began to articulate just how he differed from the sociologists. His comparative method was an historical one, comparing neighboring and contemporary societies in an effort to uncover differences and "originality," rather than comparing societies within or between social types to uncover sociological laws. Nevertheless, in time he began to soften his position. He eventually admitted that wide-ranging comparisons had a role to play in classifying societies and in shedding some light on the general characters of slavery, feudalism, and capitalism. In a more historical vein, Bloch's work contrasted with that of the later Durkheimians primarily in its insistence on an "internal" and "concrete" examination of human realities. This examination required, he argued, a close look at human motives as expressed, consciously or not, in "documents of intention." Bloch did not share Simiand's and Halbwachs' concern with identifying a distinctive field of sociology and was far more willing than they to look at human agents, the role of the individual, and other non-social factors.

In Bloch's later substantive work, he came closer than ever to adopting a sociological position. His writings on economic history consistently treated economic phenomena as social ones, and in time, he began to focus on questions relating to the general characters of differing economies that had been highlighted by Simiand, rather than questions of technology and innovations.

His writings on feudalism examined feudal society explicitly as a social type, even if one colored by a particular historical base. As he testified in a 1939 address, he came to view himself as one of those historians "accustomed to a greater or lesser degree to expressing ourselves in sociological terms."[26] His use of the terms collective consciousness, collective representations, and social substratum was, however, never strictly Durkheimian, since in Bloch's approach these terms were divorced from some of the basic aims and premises of Durkheimian sociology. Eventually, he endorsed general classifications of social types and wide-ranging comparisons and addressed broad sociological questions, but he remained primarily interested in celebrating the variety of life in hopes of cultivating a respect for the differences between people.

Bloch's clearest explanation of the difference between his form of history and Durkheimian sociology came in his unfinished *Apologie pour l'histoire*. Unlike Durkheimian sociology, he refused to make a strict separation between the individual and the social and argued that as a study of conscious beings, history must examine motives. He promoted concepts that were both flexible and linked to an historical base, even though he agreed with the sociologists on the need for abstraction and careful definition. More fundamentally, he suggested that Durkheimian sociology had relied on what was now an outdated view of science – that of a uniform model of science which sought laws of repetition. Drawing on the concepts of relativity and probability, he argued that the most the historian could hope for was multiple "causal wave trains" and an understanding of differences. Bloch never completely abandoned the search for explanation, but his form of explanation was a limited one aimed at the comprehension of change and difference rather than at prediction or even broad comparative generalization.[27]

Despite the important differences between Bloch and both the Vidalians and Durkheimians, their influence helped to separate his approach from that of his predecessors. Like Simiand, he no longer viewed history as being opposed to abstract analysis. He shared Simiand's concerns over the harmful effects of the chronological, political, and individual "idols" and also believed that the most profound causes, those most worthy of study, were social ones. Whereas the older generation of historians had drawn on a traditional historical geography, Bloch, without totally rejecting it, also drew on Vidalian geography in his search for the concrete living reality. Much like the earlier historians, he viewed geography as an auxiliary subject – insufficient in and of itself; unlike them he questioned the disciplinary independence of history. In his eyes, disciplinary boundaries needed to be broken down. He dreamt of the day when the rigid faculty system could be replaced by flexible groups of disciplines.

Bloch remained, for all these differences, very much part of the effort to professionalize French history and to put it on a surer footing. He shared with people such as Langlois and Seignobos a concern for evidence and documentary criticism, striving to be "exact." In common with Lavisse, he helped to rewrite French history, focussing on the heart of France in spite of his comparative and international perspectives. Nevertheless, like Bergson and the "generation of 1905," Bloch eventually questioned the vision of science which accompanied the push for professionalization and argued that professional education in France was overemphasized at the expense of general culture and things of the spirit. For Bloch, however, this questioning did not entail a wholesale rejection of the historians' methods or of the sociologists' concerns.

It is not an easy task to compare Bloch's positions with those of the Anglo-American historical geographers who have drawn on Bloch to justify the existence of their threatened field or to the historical sociologists who have also reacted against attempts to quantify and systematize theirs. The debates in which he was involved shared little with those sparked by the "quantitative revolution." History as practiced by Langlois and Seignobos was concerned with evidence, but that evidence was not the "data" of the quantitative social scientists, subject as it was to strict rules of textual criticism and rarely taking a numerical form. Simiand's "positive" science was also very unlike that of the model builders of the 1960s, other than the high regard both held for statistical methods and general statements.

The debates of recent times may share, nevertheless, a common cause with those in which Bloch was engaged: the effort of twentieth-century academics to identify and defend distinctive disciplines in the face of pressures both for increasing specialization and for interdisciplinary approaches. As sociology departments and geography departments continue to struggle for academic legitimacy and to be threatened at a time of increasing budget restraint, it seems appropriate to reassess the challenges which those fields hold. Such a task should involve an examination of what their particular perspectives have to offer and of the extent to which those perspectives require or justify the existence of separate disciplines.

Today, interdisciplinary approaches continue to be championed and increasingly there has been a significant "blurring of genres," but often without a careful examination of how the assumptions, methodologies, and practices of the disciplines to be joined complement or contradict each other.[28] This holds not just for interdisciplinary work of the kind that Bloch envisioned but also for the more recent attempts to incorporate the methods and findings of linguistic theory into a broad range of disciplines. Bloch's work certainly does not provide a definitive solution to these problems, but it does, at least, point to some of the questions which should be raised.

Those attempting to combine geography, sociology, and history should address the questions of whether these fields really share a common object and of what each field can contribute to an understanding of that object. For example, is the geographical contribution limited to providing some sense of the concrete living reality as Bloch appeared to imply? If the object is a social one, can it be explained or understood in terms of individual or physical phenomena? Just what is to be the role of human agency and how should one view the relationships between the individual and the social? If the concrete reality really is so important, just what is the ultimate aim of the inquiry: to celebrate variety and differences, as the historian Bloch would have it, or to uncover more general characteristics and relationships? What kinds of concepts are appropriate to such a study and to what extent should our attention be turned to an examination of both our language and that used by those whom we study?

A major focus of contemporary discussions within sociology and geography on the possible links between sociology, geography, and history has been structuration theory as espoused, perhaps most notably, by Anthony Giddens and introduced to geographical literature by Derek Gregory, Allan Pred, and others.[29] Although certainly never clearly specified, in a surprising number of ways Bloch's concerns prefigured these discussions. The issues of human agency and the relationships between the individual and the social are clearly central to these discussions. So too are questions of language and the nature of concepts and the relationships between the language of every day and every place to those of the observer. Also in common with later writers, Bloch, by the end of his life challenged both the concepts of the positivistic character of natural science and that of the unity of science without, however, completely coming to terms with the role of "laws" and "explanation." His advocacy of an historical approach to comparison is also consistent with Giddens' view that laws are intrinsically historical in character.

On the other hand, Bloch's depictions of the role of geography and the "concrete living reality" are rather different from those espoused in the structuration discussions in which participants have often drawn on "time-geography" – speaking of time-space paths, distanciation and routination. Although Bloch did refer to social geography and drew on what could be seen as social spaces, he was less concerned with the meaning and use of space than has been the case in the structuration debate – not surprisingly so given the character of geography at his time.[30] Nevertheless, Bloch's refusal to allow a separate subject for geography and his insistence on the centrality of the social subject are consistent with the arguments of Giddens and Gregory that "like time space does not exist save as a property of objects and events" and that "time-space relations are constitutive."[31]

In some ways, Bloch's work also prefigured some of the current debates within French social history so aptly described by Roger Chartier. Increasingly French historians have abandoned the territorial bases taken from human geography which were once so central to their work and have become more concerned with regularities than particularities. As such, they, like Bloch in his comments on Simiand's critique of geography, have moved closer to

Durkheimian sociology, while paradoxically often remaining interested in the exceptional. In contrast to Bloch, however, this interest in the different tends to be used as a way to approach the normal and common, rather than as an end in itself as Bloch at times seemed to imply. Also, parallels can be drawn between current challenges to the use of prefixed social categories and Bloch's criticisms of monolithic treatments of societies and his quest for an internal approach to social groups as expressed, for example, in his reviews of Halbwachs. Furthermore, later challenges to "total history" bear much in common with Bloch's problem-centered approach and his rejection of deterministic approaches, articulated among other places in his many reviews of geographical works, and with his focus on representations and the "mental," so apparent in his writings on social class. Certainly the current debate has gone into much greater depth on the role and forms of representation, levels of discourse, and the practice of power than Bloch ever contemplated, but nevertheless, in its overall thrust, his work remains surprisingly relevant and current.[32]

Bloch's work should also be of interest to those attempting to understand Vidalian geography, which continues to be used as a reference point for geographical research. Before championing Vidal as a humanist or neo-Kantian, one would do well to understand Vidal's project in the context in which it was developed and to ask whether the Vidalians ever gave satisfactory answers to the challenges they faced from both sociologists and historians, including Bloch. The Vidalian approach certainly did make a contribution to the development of the social sciences, but it is not at all clear that many later interpretations have captured the character of that contribution.[33] Although the Vidalians helped to switch attention to social and economic history and highlighted regional differences, one should not forget Vidal's defence of his field as a biological science, the reluctance of the Vidalians to focus on socially defined problems or to examine historical roots, and Vidal's many inconsistencies and ambiguities.

The equivocal character of some of Bloch's positions appears to be a reflection in part of the times in which he wrote. Trained in the Nouvelle Sorbonne and yet very much aware of the criticisms of the "generation of 1905," Bloch attempted to reconcile two very different worlds. He sought explanations and even verification and wrote of social types. At the same time, however, he advocated the use of flexible concepts, insisted on the importance of one's perspective, wrote of the poetry of human destinies and celebrated "originality." He was not a model builder, but neither was he interested solely in poetic interpretations of the particular. His attempt to bring together two such different approaches was perhaps doomed to failure. Nevertheless, it posed some very interesting questions from which we can still learn whether we choose to join those attempting to reconcile "humanistic" and "scientific" approaches or not.

Bloch's role in the debates between the sociologists and geographers was often that of a mediator in a series of discussions which lacked direct confrontations and explicit replies. Although assuring the geographers that they had much to offer, he insisted that they take the sociologists' criticisms of the regional approach and geographical explanations seriously, and he demonstrated alternative approaches to topics such as those of field systems. The sociologists were criticized in turn for their view of science, for their lack of feel for the concrete living reality, and for failing to give satisfactory and workable solutions to the problem of the relationships between the individual and the social. Bloch was not one to engage in polemics and often chose to pose questions, instead of attempting to give clear and definitive answers. The positions which he did take changed with time as demonstrated by his writings on the comparative method and his approaches to questions of economic history. He also shifted his topics going back and forth between the geographical and the sociological as he experimented, not always entirely successfully, with various approaches.

Bloch's interest was never to carve a domain for a logically independent discipline; instead, he preferred to use what he found most interesting from other fields for his own ends. As a result, he never met the challenges of the sociologists and geographers head-on. He internalized and changed some of their methods without always admitting or, at times, even realizing how his approach differed from theirs. His was a pragmatic rather than a systematic approach to the questions which he found most interesting.

Although he certainly did not resolve all the issues raised by the debates, Bloch did keep the methodological questions alive at a time when many others were content to go their own ways in their separate disciplinary domains. As social scientists and historians come to question the disciplinary divisions and identifications which became entrenched during Bloch's lifetime, his work takes on new meaning as he too found those classifications restrictive and counter-productive. In his own piecemeal way, Bloch developed a thoughtful interdisciplinary approach to what he termed man in society. Perhaps one of his most significant contributions was to address the issues raised by a changing social science, without abandoning his interest in the variety of life. His approach highlighted people's thoughts, examined the realities of their daily life, and yet attempted to place that concrete living reality within broader social and economic frameworks. Clearly, there is room for other objects and approaches, but let us hope that their authors will at least reflect on the issues raised by these debates. Despite changing interests and new techniques, the disciplinary assumptions that Bloch helped to challenge continue to underly much work in geography, sociology, and history, and those assumptions are all too easily forgotten.

Notes

Introduction

1 Marc Bloch, *Apologie pour l'histoire ou métier d'historien* (Paris: Colin, 1974), p. 166.
2 Marc Bloch, "Un symptôme social: le suicide," *AHES*, 3 (1931), 591–592.
3 Marc Bloch, "L'Ile-de-France," *RSH*, 26 (1913), 349–350.
4 Marc Bloch, review of *Der Deutsche Staat des Mittelalters* (Bd. 1), by G. von Below, *RH*, 128 (1918), 344. Cf. Marc Bloch, review of *Geldwert in der Geschichte: Ein Methodologischer Versuch*, by Andreas Walther, *RSH*, 25 (1912), 244; Marc Bloch, review of *An Introduction to the Study of Prices*, by Walter T. Layton, *RSH*, 25 (1912), 106.
5 Amongst the many examples that could be cited, note the admission by Numa Broc: "Evoquer les *'Annales'* et la *Géographie* c'est presque dire *Lucien Febvre et la géographie*, tant le maître strasbourgeois a marqué de sa personnalité ce chapitre de l'histoire parallèle de nos disciplines." Numa Broc, "Les Séductions de la nouvelle géographie," in Charles-Olivier Carbonell and Georges Livet, eds., *Au berceau des Annales* (Toulouse: Presses de l'Institut d'Etudes Politiques de Toulouse, 1983), p. 256.
6 See interesting discussion on the "interdisciplinary mirage" and on the limits of a historiography based solely on an analysis of texts in Gérard Noiriel, "Pour une approche subjectiviste du social," *AESC*, 44 (1989), 1444–1446, 1449–1451.
7 For some initial forays into the international context see Peter Schöttler, " 'Désapprendre de l'Allemagne': les *Annales* et l'histoire allemande pendant l'entre-deux guerres," in Hans Manfred Bock et al., eds., *Entre Locarno et Vichy: les relations culturelles franco-allemandes dans les années 1930* (Paris: CNRS, 1993), pp. 439–461; Peter Schöttler, "Die Annales und Osterreich in den zwanziger und dreissiger Jahren," *Osterreichische Zeitschrift für Geschichtswissenschaften,* 3 (1993), pp. 74–99; Karl Ferdinand Werner, "Marc Bloch et la recherche historique allemande," in Hartmut Atsma and André Burguière, eds., *Marc Bloch aujourd'hui: histoire comparée et sciences sociales* (Paris: EHSS, 1990), pp. 125–133; Pierre Toubert, "Preface" to Marc Bloch, *Les Caractères originaux de l'histoire rurale française* (Paris: Colin, 1988), pp. 6–13; Andrzej Feliks Grabski, "Marc Bloch Przed 'Annales'," *Dzieje Najnowsze*, 13 (1981), 129–139.
8 See Noiriel, "Pour une approche," pp. 1447–1448.

9 See, for example, Marc Bloch, "Le Salaire et les fluctuations économiques à longue période," *RH*, 173 (1934), 2.

1 Marc Bloch and the "Université"

1 See interesting discussion in Eugen Weber, *France, Fin de Siècle* (Boston: Belnap/Harvard, 1986) and Theodore Zeldin, *France 1848–1945: Intellect and Pride* (Oxford University Press, 1980), pp. 29–85. Also, see the contemporary novel, Maurice Barrès, *Les Déracinés* (Paris: E. Fasquelle, 1897). For one of the most comprehensive discussions of the regionalist movement see Jean Charles-Brun, *Le Régionalisme* (Paris: Blond, 1911).
2 Richard D. Mandell, *Paris 1900, the Great World's Fair* (University of Toronto Press, 1967).
3 Marc Bloch, *Apologie pour l'histoire ou métier d'historien* (Paris: Colin, 1974), p. 151.
4 See, for example, Albert Thibaudet, *La République des professeurs* (Paris: Grasset, 1927), pp. 25–29. It is important, however, not to exaggerate the extent of the teachers' influence. On the teachers as a group apart, see, Zeldin, *France 1848–1945*, pp. 302–315.
5 AN 318 MI 1, no. 95, letter of Marc Bloch to Lucien Febvre, Apr. 2, 1933.
6 AN F^{17} 22468.
7 Raoul Blanchard, *Ma jeunesse sous l'aile de Péguy* (Paris: Fayard, 1961), p. 224. See also, Hubert Bourgin, *De Jaurès à Léon Blum: L'Ecole Normale et la politique* (Paris: Librairie Arthème Fayard, 1938), pp. 30–31, and Paul Léon, *Du Palais Royal au Palais Bourbon: souvenirs* (Paris: Editions Albin Michel, 1947), p. 48.
8 Marc Bloch, *Les Rois thaumaturges* (Paris: Gallimard, 1983), p. xl (originally published in 1924 by the Faculty of Letters of the University of Strasbourg).
9 Jérome Carcopino, "Gustave Bloch," ENS, Association Amicale des Anciens Elèves de l'Ecole Normale Supérieure, *Annuaire*, 1925, 2, p. 108. In the original "mediatif" is printed instead of "meditatif."
10 Lucien Febvre, "Marc Bloch et Strasbourg: Souvenirs d'une grande histoire," in *Mémorial des années 1939–1945* (Paris: Belles Lettres, 1947), Publications de la Faculté des Lettres de l'Université de Strasbourg, fasc. 103, p. 172.
11 On Louis's death see Bertrand Müller, "Introduction," to Marc Bloch and Lucien Febvre, *Correspondance*, I (*La Naissance des Annales*), ed. Bertrand Müller (Paris: Fayard, 1994), p. xii; Carole Fink, *Marc Bloch: A Life in History* (Cambridge University Press, 1989), pp. 14–17.
12 Fink, *Marc Bloch*, pp. 14–17.
13 Marc Bloch, *Les Rois thaumaturges* (Paris: Gallimard, 1983), p. xli. In 1920, he dedicated *Rois et serfs* to "Mon père, son élève": Marc Bloch, *Rois et serfs: un chapitre d'histoire capétienne* (Paris: Honoré Champion, 1920).
14 Zeldin, *France 1848–1945*, p. 276.
15 Fink, *Marc Bloch*, p. 24.
16 AN 61 AJ 233.
17 Howard F. Andrews, "Paul Vidal de la Blache and the Concours d'Agrégation of 1866," *Canadian Geographer*, 31: 1 (1987), 16–17.
18 Fink, *Marc Bloch*, p. 27.

19 Howard F. Andrews, "A Note on Agrégation during the July Monarchy and the IInd Republic (1830–51)," mimeo, University of Toronto, Department of Geography, p. 3.

20 Georges Chabot, "La Genèse de l'agrégation de géographie," *AG*, 85 (1976), 333–340; Numa Broc, "L'Etablissement de la géographie en France: diffusions, institutions, projets (1870–1890)," *AG*, 83 (1974), 561–562. On the well developed institutional support for history, see Madeleine Rebérioux, "Histoire, historiens et dreyfusisme," *RH*, 518 (1976), 410–411.

21 Claude Digeon, *La Crise allemande de la pensée française (1870–1914)* (Paris: PUF, 1959), pp. 375–383.

22 On Bloch's year in Germany, see Peter Schöttler, "'Désapprendre de l'Allemagne': les *Annales* et l'histoire allemande pendant l'entre-deux guerres," in Hans Manfred Bock et al., eds., *Entre Locarno et Vichy: les relations culturelles franco-allemandes dans les années 1930* (Paris: CNRS, 1993), pp. 441–443; Pierre Toubert, "Preface," to Marc Bloch, *Les Caractères originaux de l'histoire rurale française* (Paris: Colin, 1988), pp. 7–9.

23 When advising his son Etienne on his future, Marc Bloch noted the advantages that his position at the Fondation Thiers had given him. See letter of Marc Bloch to Etienne Bloch, Apr. 24, 1940 in François Bédarida and Denis Peschanski, eds., *Marc Bloch à Etienne Bloch: lettres de la drôle de guerre* (Paris: Cahiers de l'Institut d'Histoire du Temps Présent, 1991), p. 84.

24 Marc Bloch, *Souvenirs de guerre, 1914–1915* (Paris: Colin, 1969); Carole Fink, "Introduction: Marc Bloch and World War I," in her translation, Marc Bloch, *Memoirs of War, 1914–15* (Ithaca: Cornell University Press, 1980); Fink, *Marc Bloch*, pp. 54–78.

25 AN 318 MI 1: no. 490–494, letter of Marc Bloch to Georges Davy, Sept. 16, 1917; no. 497, letter of Gustave Bloch to Georges Davy, Jan. 27, 1918. AN AJ[16] 5876.

26 AN 318 MI, 1, no. 281, letter of Marc Bloch to Lucien Febvre, Oct. 8, 1939. See also Marc Bloch, *Strange Defeat: A Statement of Evidence Written in 1940* (New York: Norton, 1968), pp. 171–173. On the political detachment of many French intellectuals in the 1930s, see François Dosse, *L'Histoire en miettes: des "Annales" à la "nouvelle histoire"* (Paris: Editions la Découverte, 1987), p. 17.

27 Febvre, "Marc Bloch et Strasbourg," p. 175; Lucien Febvre, "De l'histoire au martyre: Marc Bloch 1886–1944," *AHS*, 7 (1945), p. 2. In the *Livret Guide d'Etudiant* of the Faculty of Letters for 1923–1924, and 1924–1926, Bloch is listed at two addresses, first, 48 bis, Allée de la Robertsau, and second, 19 Avenue de la Liberté; Febvre is listed at 5, Allée de la Robertsau; and Halbwachs at 48 bis, Allée de la Robertsau. In 1931, Bloch moved to 59, Allée de la Robertsau. See Müller, "Introduction," p. xxxi. On the residential patterns of the Strasbourg professors, see Jean-Luc Pinol, "Itinéraires résidentiels des universitaires strasbourgeois," *Les Annales de la Recherche Urbaine*, 62–63 (1994), 157–168.

28 Philippe Dollinger, "Marc Bloch et Lucien Febvre à Strasbourg (Quelques souvenirs d'un de leurs étudiants)," in Charles-Olivier Carbonell and Georges Livet, eds., *Au berceau des Annales* (Toulouse: Presses de l'Institut d'Etudes Politiques de Toulouse, 1983), p. 65. (Actes du Colloque de Strasbourg, Oct. 11–13, 1979.)

29 Robert Boutrouche, "Marc Bloch vu par ses élèves," in *Mémorial des années*, pp, 196, 200. See also Carole Fink's discussion of the American graduate student

William Mendell Newman, Fink, *Marc Bloch*, p. 167. See also Lucien Febvre, "Marc Bloch," in *Architects and Craftsmen in History: Festschrift für Abbott Payson Usher* (Tübingen: J.C.B. Mohr, 1956), p. 89; On Bloch's relationships to his students see Henri Brunschwig, "Vingt ans après (1964): souvenirs sur Marc Bloch," in Marc Michel, ed., *Etudes Africaines offerts à Henri Brunschwig* (Paris: EHESS, 1982), p. xiv; François-Georges Pariset, "Du bon travail à Strasbourg. . . Note sur la Réforme," in Carbonell and Livet, *Au berceau*, p. 68; see also Etienne's Bloch recollections, AN 318 MI 1, no. 409.

30 Christian Pfister, "La Première Année de la nouvelle Université Française de Strasbourg (1918–1919)," *RIE*, 68 (1919), 333.

31 John Craig, *Scholarship and Nation Building: The Universities of Strasbourg and Alsatian Society (1870–1939)* (University of Chicago Press, 1984), pp. 312, 238–289.

32 AN 318 MI 1, no. 185–186, letter of Marc Bloch to Lucien Febvre, Mar. 30, 1936.

33 Marc Bloch, *Strange Defeat*, p. 152. Cf. Marc Bloch, "Une conscience de classe: les paysans de Norvège," *AHES*, 3 (1931), 282. There Bloch wrote of Halvdan Koht's "rigidité un peu artificielle" stemming from his "convictions marxistes."

34 Fink, *Marc Bloch*, pp. 101, 181.

35 AN 318 MI I, no. 200–201, letter of Marc Bloch to Lucien Febvre, Sept. 20, 1937.

36 Charles-Edmund Perrin, "L'Oeuvre historique de Marc Bloch," *RH*, 199 (1948), 175; Febvre, "Marc Bloch et Strasbourg," 172; Brunschwig, "Vingt ans après," p. xiv; recollections of Etienne Bloch, AN 318 MI 1, no. 410; letter of Etienne Bloch to Fernand Braudel, Aug. 10, 1962, in AN 318 MI 1, no. 383; Etienne Bloch, "Marc Bloch. Souvenirs et réflexions d'un fils sur son père," in Hartmut Atsma and André Burguière, *Marc Bloch aujourd'hui: histoire comparé et sciences sociales* (Paris: EHESS, 1990) pp. 26–27.

37 Giuliana Gemelli, "Communauté et stratégies institutionelles: Henri Berr et la Fondation du Centre International de Synthèse," *RS*, 4th ser., 2 (Apr.-June, 1987), 235–236, 238.

38 Fink, *Marc Bloch*, pp. 97, 87. Although Bloch's mother maintained her home in Marlotte, she made extended visits to Strasbourg (e.g. winter 1924). See, for example, AN 318 MI 1, no. 511, letter of Marc Bloch to Georges Davy, Mar. 16, 1924.

39 Etienne Bloch, "Ce que mon père nous racontait," *Nouvel Observateur*, 981 (Aug. 28, 1983), pp. 68–69; AN 318 MI 1, no. 410 (recollections of Etienne Bloch); Etienne Bloch, "Marc Bloch. Souvenirs et réflexions," pp. 23–24.

40 Cf. Dosse on marginality as part of the *mythe fondateur* of the Annales: Dosse, *L'Histoire*, p. 39. See also Bertrand Müller, "Introduction," pp. v-vii.

41 Charles-Olivier Carbonell, "Les Professeurs d'histoire de l'enseignement supérieure au début du XXᵉ," in Carbonell and Livet, eds., *Au berceau*, p. 93; see also Olivier Dumoulin, "Changer l'histoire. Marché universitaire et innovation intellectuelle à l'époque de Marc Bloch," in Atsma and Burguière, *Marc Bloch aujourd'hui*, pp. 87–104.

42 See Burguière on the novelty of the program of the *Annales* as more in the way the program was affirmed than in the program itself: André Burguière, "L'Histoire d'une histoire: la naissance des Annales," *AESC*, 11 (1979), 1350. See also Dosse's argument that *l'histoire positiviste* attacked by the *Annales* was not a particularly dangerous adversary since it was already dead: Dosse, *L'Histoire*, pp. 52–53.

43 Alice Gérard, "A l'origine du combat des Annales: Positivisme historique et système universitaire," in Carbonell and Livet, *Au berceau,* p. 85; Carbonell, pp. 100–103. See also Antoine Prost, "Seignobos revisité," *Vingtième Siècle,* 43 (1994), 112–117.

44 On the tension and resulting ambiguities stemming from geography's start as a scholastic discipline see Catherine Rhein, "La Géographie, discipline scolaire et/ou science social?" *RFS,* 23 (1982), 223–251; Isabele Lefort, *La Lettre et l'esprit: géographie scolaire et géographie savante en France* (Paris: CNRS, 1992). On geography and agrégation, see Marie-Paule Maret and Philippe Pinchemel, "Evolution des questions de géographie aux cours d'agrégation des origines à 1914: contribution à l'histoire de la pensée géographique," *La Pensée géographique française* (Saint-Brieuc: P.U. de Bretagne, 1972), pp. 77–86.

45 See, for example, Emile Durkheim, "L'Enseignement philosophique et l'agrégation de philosophie," *Revue Philosophique de la France et de l'Etranger,* 39 (1885), 121–147.

46 Howard F. Andrews, "The early life of Paul Vidal de la Blache and the makings of modern geography," *Transactions of the Institute of British Geography,* n.s. 11 (1986), 179; Victor Karady, "Stratégies de réussite et modes et faire-valoir de la sociologie chez les durkheimiens," *RFS,* 20 (1979), 56–57.

47 On post-war internationalism, see Dosse, *L'Histoire,* p. 15.

48 On the image of France as a hexagon, see Lefort, *La Lettre,* pp. 172–173.

49 See Marc Bloch and Febvre, *Correspondance,* I, pp. 259–260, 263–267 (letters of Lucien Febvre to Marc Bloch, Dec. 6 and 31, 1930 with notes by Bertrand Müller). The salaries at the Ecole Pratique des Hautes Etudes were about half those of Strasbourg. On this institution, see Terry N. Clark, *Prophets and Patrons: The French University and the Emergence of Social Science* (Cambridge, MA: Harvard, 1973), pp. 42–51.

50 AN 318 MI 1, no. 188, letter of Marc Bloch to Lucien Febvre, Apr. 19, 1936.

51 In 1934, Febvre contemplated the interesting possibility of Bloch, Simiand, and himself giving a joint series of lectures: AN 318 MI 2, no. 445–446, letter of Lucien Febvre to Marc Bloch, n.d., c. Mar., 1934.

52 AN 318 MI 1, no. 240–241, no. 242–244, letters of Marc Bloch to Lucien Febvre, Dec. 5 and 7, 1938; see also no. 458–460, letter of Marc Bloch to Célestin Bouglé, May 18, 1939, and AN 318 MI 2 no. 461–463, letter of Lucien Febvre to Marc Bloch, n.d., c. Mar. 1934.

53 Fink, "Introduction: Marc Bloch and World War I," p. 61; Georges Friedmann, "Au dela de 'l'engagement': Marc Bloch, Jean Cavaillès," *Europe,* 24: 10 (Oct. 1, 1946), 25.

54 Georges Altman, "Notre 'Narbonne' de la Résistance," *Cahiers Politiques,* 8 (Mar. 1945), 3; Febvre, "Marc Bloch et Strasbourg," p. 185; Fink, *Marc Bloch,* p. 199.

55 AN 318 MI, 1, no. 281, letter of Marc Bloch to Lucien Febvre, July 6, 1940.

56 See for example, Etienne Bloch, "Ce que mon père," p. 70.

57 AN 318 MI 1, no. 254, letter of Marc Bloch to Lucien Febvre, Oct. 8, 1939; see also AN 318 MI 1, no. 250–251, letter of Marc Bloch to Lucien Febvre, Sept. 17, 1939.

58 AN 318 MI 1, no. 287, 290, 291, letters of Marc Bloch to Lucien Febvre, Sept. 8 and 26, 1940; AN AJ[16] 5876.

59 Fink, *Marc Bloch,* pp. 249–266; Peter M. Rutkoff and William B. Scott, "Letters to

America: the correspondance of Marc Bloch, 1940–1941," *French Historical Studies*, 12: 2 (1981), 277–303.

60 Fink, *Marc Bloch*, pp. 264–292. According to Fink, Fliche, nursing resentment from a critical review by Bloch, not only attempted to obstruct Bloch's nomination, but also in 1942 did not include Bloch's work in his annual report on the accomplishments of the faculty.

61 The best source for Bloch's resistance activities and subsequent capture is Fink, *Marc Bloch*, pp. 293–324.

62 Febvre, "Marc Bloch et Strasbourg," p. 188.

63 Re ministerial dreams see Fink, *Marc Bloch*, p. 302 and Eugen Weber, "About Marc Bloch," *American Scholar*, 51: 1 (Winter 1981/1982), 79. See also, Dominique Veillon, *Le Franc-Tireur: un journal clandestin, un mouvement de Résistance, 1940–1944* (Paris: Flammarion, 1977), p. 340.

According to Fink and Veillon, Bloch was named editor-in-chief of the *Cahiers Politiques*, but François de Menthon and Lucien Febvre both reported that he declined that position. Fink, *Marc Bloch*, p. 304; François de Menthon, "A la mémoire de Marc Bloch," *Cahiers Politiques*, 8 (1945), 1; Febvre, "Marc Bloch et Strasbourg," p. 188.

64 Marc Bloch, "Sur les programmes d'histoire de l'enseignement secondaire," *Bulletin de la Société des Professeurs d'Histoire et de Géographie de l'Enseignement Public*, 12: 29 (1921), 15–17.

65 Marc Bloch and Lucien Febvre, "A propos d'un concours," *AHES*, 6 (1934), 265–266. On the drafting of this article see EB: letter of Marc Bloch to Lucien Febvre, Dec. 27, 1933. Marc Bloch and Lucien Febvre, "Le Problème de l'agrégation," *AHES*, 9 (1937), 115–117.

66 Marc Bloch, "Notes pour une révolution de l'enseignement," *Cahiers Politiques*, Aug. 3, 1943, p. 21.

67 Ibid., p. 24.

2 Marc Bloch's training as a *normalien*

1 Marc Bloch, *Apologie pour l'histoire ou métier d'un historien* (Paris: Colin, 1974), p. 151.

2 A. Aulard, "La Transformation de l'Ecole Normale Supérieure," *Revue Bleue*, 4th ser., 19 (Jan.–June, 1903), 53; Marc Bloch, "Jean Jaurès," *AHES*, 2 (1930), 629–630; Carole Fink, "Introduction: Marc Bloch and World War I," in her translation, Marc Bloch, *Memoirs of War, 1914–1915* (Ithaca: Cornell University Press, 1980), pp. 26–30; Robert Smith, "L'Atmosphère politique à l'Ecole Normale Supérieure à la fin du XIXe siècle," *Revue d'Histoire Moderne et Contemporaine*, 20 (1973), 248–268.

3 Ernest Lavisse, "Ecole Normale Supérieure: Séance du mercredi 23 novembre 1904," *RIE*, 48 (1904), 490–492. During the height of the Dreyfus Affair, Lavisse himself had tried to play a moderating role. See Madeleine Rebérioux, *La République radicale, 1898–1914* (Paris: Seuil, 1975), p. 26. On Lavisse's tardy republicanism see François Dosse, *L'Histoire en miettes: des "Annales" à la "nouvelle histoire"* (Paris: Editions la Découverte, 1987), p. 34.

4 In part this was due to a transition period; second and third year students were to

follow the old régime. However, even for Bloch's year many of the courses were taken at the ENS. In Bloch's class, for example, students in history and geography were required to take three *enseignements généraux* at the Faculty and one *conférence pratique* at the ENS for each of ancient, medieval, and modern history, and geography. Although the *conférences pratiques* were open to students from the Faculty, only one such student enrolled in history. As similar *cours pratiques* were also held at the Faculty of Letters, faculty students would have needed considerable ambition and courage to try to penetrate those of the ENS. See AN 61 AJ 186; Ernest Lavisse, "L'Ecole Normale Supérieure," in Académie de Paris, *Rapports présentés au conseil académique sur les travaux et les actes des établissements d'enseignement supérieur pendant l'année 1904–1905* (Melun: Imprimerie administrative, 1906), pp. 117–121; Pierre Jeannin, *Ecole Normale Supérieure: livre d'or* (Paris: Office Français de Diffusion Artistique et Littéraire, 1963), pp. 116, 120.

5 As a fairly comprehensive source of the occupations, one can use the annotated copy of the 1939 *Annuaire* of the Association Amicale de Secours des Anciens Elèves of the ENS found in AN 61 AJ 207. Of those still living in 1939, the numbers listed as occupying positions other than teaching and educational administration (for the most part, political, governmental, and literary positions) were 1 out of 11 (1902), 2 out of 16 (1903), 1 out of 15 (1904), 3 out of 22 (1905), 6 out of 27 (1906). By far the largest numbers held active teaching positions in the Faculties of Letters, lycées, and collèges. This tabulation excludes those for whom no occupation was listed, i.e. those not members of the association – in all 1 for 1902, 1 for 1903, 1 for 1905, and 2 for 1906. Despite the concerns over *normaliens* pursuing other careers, these figures differed little from those in the late nineteenth century. See Robert J. Smith, *The Ecole Normale Supérieure and the Third Republic* (Albany: State University of New York, 1982), pp. 50–51.

6 On the administrative reforms, see George Weisz, *The Emergence of Modern Universities in France* (Princeton University Press, 1983); William Arthur Bruneau, "French universities and faculties, 1870–1902," Ph.D., Dept. of Educational Theory, University of Toronto, 1977; Ferdinand Lot, "De la situation faite à l'enseignement supérieure en France," *Cahiers de la Quinzaine*, ser. (1905–6) 9, pp. 1–96, and 11, pp. 109–237.

7 On Durkheim's appointment and his opponents, see Steven Lukes, *Emile Durkheim, His Life and Work: A Historical and Critical Study* (Harmondsworth: Penguin, 1973), pp. 359–378.

8 George Weisz, "Republican ideology and the social sciences: the Durkheimians and the history of social economy at the Sorbonne," in Philippe Besnard, ed., *The Sociological Domain: The Durkheimians and the Founding of French Sociology* (Cambridge University Press and Paris: Editions de la Maison des Sciences de l'Homme, 1983), p. 106. In 1904, the Ecole des Hautes Etudes Sociales sponsored a debate between Durkheim and Tarde. See Terry N. Clark, *Prophets and Patrons* (Cambridge, MA: Harvard University Press, 1973), p. 160; Lukes, *Emile Durkheim*, pp. 312–313.

9 Among the French participants in Worms' projects were Tarde, Espinas, Gide, Cheysson, Monod, and Ribot; the foreigners included such notables as Simmel, Tönnies, Veblen, Giddings, and Ward.

10 See Université de Paris, *Livret de l'étudiant, 1907–1908* (Paris: Bureau des

Renseignements à la Sorbonne, n.d.); *Minerva*, 1907–1908; and Besnard, ed., Introduction to *The Sociological Domain*, pp. 3–4. For more general discussions see Clark, *Prophets*, pp. 162–195 and Victor Karady, "Durkheim, les sciences sociales et l'Université: bilan d'un semi-échec," *RFS*, 17 (1976), pp. 311.

11 Victor Karady, "The Durkheimians in Academe: A reconsideration," in Besnard, ed., *The Sociological Domain*, p. 74.

12 Marc Bloch, "Les Annales sociologiques," *AHES*, 7 (1935), 393; Marc Bloch, "Le Salaire et les fluctuations économiques à longue période," *RH*, 173 (1934), 2.

13 Karady has argued that geography was institutionalized much more quickly than the other social sciences. However, Catherine Rhein has noted that in fact the process was rather slow. Vidal took far longer than Durkheim to gain a Sorbonne chair, and geography did not gain a separate *agrégation* until 1944. Karady, "Durkheim," pp. 275–277; Catherine Rhein, "La Géographie, discipline scolaire et/ou science sociale (1860–1920)," *RFS*, 18 (1982), 240.

14 See the synoptic tables in Université de Paris, *Livret de l'étudiant, 1905–1906* (Melun: Imprimerie Administrative, 1905), pp. 233, 236. For changes, see Université de Paris, *Livret de l'étudiant, 1907–1908*; *Minerva*, 1907–1908. Bérard had been promoted from a *maître de conference* to a *maître adjoint* at the Ecole Pratique des Hautes Etudes, and at the Ecole Libre des Sciences Politiques, Leblond was replaced by Cloarec.

15 Vincent Berdoulay, *La Formation de l'école française de géographie* (Paris: B.N., 1981), pp. 102–103; Vincent Berdoulay, "Louis-Auguste Himly, 1823–1906," *Geographers: Biobibliographical Studies*, I, (Salem, NH: Mansell, 1977), pp. 43–47. Se also, André-Louis Sanguin, *Vidal de la Blache, 1845–1918: un génie de géographie* (Paris: Belin, 1993), p. 260.

16 In 1872, Vidal was a *chargé de cours* and in 1875 became a titled professor.

17 Howard F. Andrews, "The early career of Paul Vidal de la Blache and the makings of modern geography," *Transactions, Institute of British Geographers*, n.s. 11 (1986), 174–182; Howard F. Andrews, "Les Premiers cours de géographie de Paul Vidal de la Blache à Nancy (1873–1877)", *AG*, 95 (1986), 341–361; Sanguin, *Vidal de la Blache*, pp. 99–118.

18 In 1895 following a rift from Vidal, Dubois was replaced as co-directeur by Emmanuel de Margerie and Lucien Gallois. Soon thereafter, the colonial theme became less evident. See Sanguin, *Vidal de la Blache*, pp. 130–131.

19 Les Directeurs (P. Vidal de la Blache and Marcel Dubois), "Avis au lecteur," *AG*, 1 (1891), ii. See also Michel Chevalier, "Géographie ouverte et géographie fermée: les premières années des Annales de Géographie," in Paul Claval, *Autour de Vidal de la Blache* (Paris: CNRS, 1993). pp. 133–134.

20 Lucien Febvre, "Marc Bloch et Strasbourg: souvenirs d'une grande histoire" in *Mémorial des années 1939–1945* (Paris: Belles Lettres, 1947), Publications de la Faculté des Lettres de l'Université de Strasbourg, fac. 103, p. 176. Cf. Paul Léon, "Délaissant les historiens, c'est vers la géographie que je me sentais attiré. Elle m'apparaissait comme une vision neuve des choses, une manière insoupçonnée de considerer l'univers, une féconde collaboration des sciences physiques, naturelles, humaines." Paul Léon, *Du Palais Royale au Palais Bourbon: souvenirs* (Paris: Albin Michel, 1947), pp. 55–56.

21 In this tabulation Pfister and Monod were counted as only one position as were

Seignobos and Lavisse (see Table 2). For the early positions in history at Paris, see A. Himly, "Livret de la Faculté des Lettres de Paris," *RIE*, 5 (1883), 269–285.

22 New posts included a professorship for "Histoire du Christianisme dans les temps modernes" and at the level of *chargé de cours:* "Histoire du christianisme," "Histoire de la littérature et des idées chrétiennes depuis le XVIe," and "Histoire coloniale." The proliferation of chairs on the history of Christianity may have reflected a concern that religious studies be put on a more objective basis – a concern shared by many of the Durkheimians. Debidour who held the new professorship was known for his strong anti-clerical beliefs, and Guignebert, who taught the course on the history of Christianity, strove to develop a rationalist interpretation of *le fait chrétien*. Université de Paris, *Livret de l'étudiant, 1907–1908*; *Minerva, 1907–1908*; pp. 928–929.

23 William R. Keylor, *Academy and Community* (Cambridge, MA: Harvard University Press, 1975), pp. 39 and 60ff. Emile Durkheim, *Education pédagogique en France*, II (*De la Renaissance à nos jours*) (Paris: Alcan, 1938), pp. 203–204.

24 Pierre Leguay, *La Sorbonne* (Paris: Grasset, 1910), p. 36; Louis Liard, *Universités et facultés* (Paris: Colin, 1890), pp. 108–109.

25 In the 1860s, realizing the difficulty of opening up the Sorbonne to more rigorous methodology, Victor Duruy founded the Ecole Pratique des Hautes Etudes in which practical experience in historical criticism was given in seminars, following the German example.

26 Charles-Victor Langlois and Charles Seignobos, *Introduction aux études historiques* (Paris: Hachette, 1898).

27 Keylor, *Academy*, pp. 75–82.

28 Ibid. pp. 82–86, 172; Weisz, *The Emergence*, pp. 215–216.

29 G. Monod and G. Fagniez, "Avant-propos," *RH*, 1 (1876), 1. *La Revolution Française* appeared in 1881 and soon led to the foundation of the Société de l'Histoire de la Revolution Française which assumed control over its publication. Later publications included the *Revue Bossuet* (1900), *Revue Bourdaloue* (1902), *Revue des Etudes Rabelaisiennes* (1903), and the *Bulletin de la Société d'Histoire de la Révolution* (1904).

30 "Première séance provisoire du 22 juillet 1901," *BSHM*, 1 (July 1901), 1. Keylor, *Academy*, p. 174.

31 For a list of the "mémoires inscrits" for the session of June 1906, see AN 61 AJ 186.

32 AN 318 MI 1 no. 431, letter of Christian Pfister to Emile Boutroux, Apr. 29, 1908.

33 For a list of the *mémoire* topics (excluding those of the ENS) which were *admis* see Achille Luchaire, "Le Nouveau Diplôme d'Etudes Supérieures d'Histoire et de Géographie dans l'Université de France: les examens et le diplôme d'histoire et de géographie en Sorbonne," *RIE*, 34 (July–Dec., 1897), 199–200. On the medieval training of the majority of the French historians at the turn of the century, see Charles-Olivier Carbonell, "Les Professeurs d'histoire de l'enseignement supérieur en France au début du XXe siècle," in Charles-Olivier Carbonell and Georges Livet, eds., *Au berceau des Annales* (Toulouse: Presses de l'Institut d'Etudes Politiques de Toulouse, 1983), pp. 97–98.

34 Charles-Olivier Carbonell, *Histoire et historiens: une mutation idéologique des historiens français, 1865–1885* (Toulouse: Privat, 1976), p. 268. Cf. Lot's study of the DES in history for the years 1907, 1908, and 1909, which also found an increase

in modern history at the expense of medieval, described in Keylor, *Academy,* p. 172.

35 AN 61 AJ 86. The preceding class of twenty-five also included three (Georges Gelley, Jean Reynier, and Antoine Bianconi) who would later contribute to the *Année Sociologique,* and the following class of twenty-eight included two others (Georges Davy and Henri Jeanmaire). See also testimony in Marcel Mauss, *Oeuvres* (Paris: Editions de Minuit, 1969), III, p. 485 and Marcel Mauss, "In memoriam: L'oeuvre inédit de Durkheim et de ses collaborateurs," *AS,* n.s. 1 (1923–1924), pp. 19–20.

36 See lists of *promotions* in Association Amicale de Secours des Anciens Elèves de l'ENS, *Annuaire.* Cf. Howard F. Andrews, "The Durkheimians and human geography: some contextural problems in the sociology of knowledge," *Transactions of the Institute of British Geographers,* n.s. (1984), figure 2, p. 321.

37 Ch. V. Langlois, "Agrégation d'Histoire et de Géographie, Concours de 1908," *RU,* 17: 2 (1908), 277–293. This situation raised considerable protest just as Bloch entered the ENS. See the responses received by Berr in his survey of historians regarding "L'Enseignement supérieure en histoire" in *RSH,* 9 (1904). Note especially those of Albert Dufourcq, p. 33, P. Boissonnade, p. 162, and Desdevises du Dezert, p. 170.

The shift in the question for geography from those of *la géographie historique* to those of Vidalian geography happened very rapidly following a reorganization of *agrégation* in 1894. See Marie-Paule Maret and Philipe Pinchemel, "Evolution des questions de géographie aux concours d'agrégation des origines à 1914: contribution à l'histoire de la pensée géographique," in *La Pensée géographique française contemporaine* (Saint Brieuc: P.U. de Bretagne, 1972), pp. 77–86.

38 Louis Liard, *L'Université de Paris* (Paris: Renouard, 1909), I (La Vieille Université, la Nouvelle Université, la Nouvelle Sorbonne), pp. 90–91.

39 Re Lavisse, see Louis Liard, *L'Enseignement supérieur en France,* II (Paris: Colin, 1894), p. 352. Bloch and Lot remained in touch for years. See for example letters from Lot to Bloch in AN AB[XIX] 3851, Apr. 7, 1924 and in AN AB[XIX] 3803, Oct. 1, 1932.

40 Lot, "De la situation faite," vol. 11, pp. 136–138. "PCN" was a certificate of *sciences physiques, chimiques et naturelles* required for entrance to the Faculties of Medicine.

41 Henri Berr, "Sur notre programme," *RSH,* 1 (1900), 1–8. Berr anticipated and received contributions from a variety of fields. By 1904, contributors included the philosophers Bernedetto Croce and Emile Boutroux; the sociologists Durkheim, Henri Hubert, and Simiand; the geographers Vidal de la Blache and Pierre Foncin; as well as such historians as Karl Lamprecht, Paul Lacombe, Paul Mantoux, Charles-Victor Langlois, Henri Sée, and Jacques Flasch.

42 Lucien Febvre, "De la *Revue de Synthèse* aux *Annales*: Henri Berr ou un demi-siècle de travail au service de l'histoire," *AESC,* 7 (1952), 290.

43 Liard, *Universités et facultés,* pp. 151–152.

44 The French universities had been officially disbanded in 1793. In 1808, Napoleon established a centralized state-run "Université" which incorporated teaching posts from the primary to the faculty levels. The remaining faculties, now state controlled and no longer part of corporate university bodies, served mainly as

examining bodies for the *baccalauréat* and and the *licence*. Following the defeat of France in the Franco-Prussian war in 1870, a movement to reform the educational system and re-establish the universities gained strength. The universities regained corporate identity in 1896. See, for example, Weisz, *The Emergence*, pp. 134–161.

45 Lot, "De la situation faite," vol. 9, p. 74; vol. 11, p. 109. Jean Charles-Brun, *Le Régionalisme* (Paris: Blond, 1911), pp. 150–155.

46 In part these measures were meant to undermine the activities of Arcisse de Caumont, a *savant* from Caen who fought any official association with the Ministry of Public Instruction. See Robert Fox, "Learning, politics, and polite culture in provincial France: the Sociétés Savantes in the nineteenth century," *Historical Reflections*, 7 (1980), 549–552.

47 On the *sociétés savants,* see also Robert Fox, "The savant confronts his peers: scientific societies in France, 1815–1914," in Robert Fox and George Weisz, eds., *The Organization of Science and Technology in France, 1808–1914* (Cambridge University Press, 1980), pp. 241–282.

48 Maurice Dumoulin, "Du groupement des sociétés savantes en vue de travaux communs," *Revue des Etudes Historiques* (1899), 82; "Séance de clôture," *Comité des Travaux Historiques et Scientifiques: Bulletin Historique et Philologique* (1900), 188.

49 "Séance du 31 octobre," *BSHM*, (Nov. 1901), 5–7; "Séance du 2 janvier 1902," *BSHM*, 4 (Jan. 1902), 18–20; Henri Berr, "Les Régions de la France," *RSH*, 6 (1903), 169–171. See also Lanson's contributions to these discussions in Gustave Lanson, "Programme d'études sur l'histoire provinciale de la vie littéraire en France," *Revue d'Histoire Moderne et Contemporaine*, 4 (1902–1903), 445–464 (address given on Feb. 7 to the Société d'Histoire Moderne).

50 Victor Karady, "Recherches sur la morphologie du corps universitaire littéraire sous la Troisième République," *Le Mouvement Social*, 96 (July–Sept., 1976), 47–79; Victor Karady, "Normaliens et autres enseignants à la Belle Epoque," *RFS*, 13 (1972), 35–58.

51 Maurice Dumoulin, "Choses à faire," *RSH*, 3 (1901), 298.

52 This term was taken from H. Stuart Hughes, *Consciousness and Society* (New York: Vintage, 1977), pp. 337–344. See also Robert Wohl, *The Generation of 1914* (London: Weidenfeld and Nicolson, 1980), esp. pp. 5–41; Phyllis H. Stock, "Students versus the university in pre-World War Paris," *French Historical Studies*, 7, (1971); Clark, *Prophets,* pp. 190–194.

53 Note the description by Massis of Bergson's course in Henri Massis, *Evocations: souvenirs, 1905–1911* (Paris: Plon, 1931), pp. 90–96. See also, Hughes, *Consciousness*, p. 341.

54 Massis, *Evocations,* p. 9. See also Jean Madrian, "Chronique Biographique," in *Itinéraires*, 49 (Jan., 1961), 189.

55 These articles are collected in Agathon, *L'Esprit de la Nouvelle Sorbonne* (Paris: Mercure de France, 1911).

56 Massis, *Evocations,* pp. 63–83.

57 Reprinted in Agathon, *L'Esprit de la Nouvelle Sorbonne*, p. 364.

58 René Benjamin, *La Farce de la Nouvelle Sorbonne* (Paris: M. Rivière, 1911); Pierre Lasserre, *La Doctrine officielle de l'Université* (Paris: Mercure de France, 1913).

59 Agathon, *Les Jeunes gens d'aujourd'hui* (Paris, Plon, s.d. c. 1919), pp. 68, 99, 170–174.

3 History under attack

1 Marc Bloch, "Méthodologie historique," *Rivista di Storia della Storiografia Moderna*, 9, 2–3 (1988), 155.

2 See discussion in Claude Digeon, *La Crise allemande de la pensée française* (Paris: PUF, 1959), p. 376.

3 Georg Iggers, *New Directions in European Historiography* (Middletown: Wesleyan University Press, 1975), pp. 27–31.

4 See discussions in Madeleine Rebérioux, "Le Débat de 1903: historiens et sociologues," in Charles-Olivier Carbonell and Georges Livet, eds., *Au berceau des Annales* (Toulouse: Presses de l'Institut d'Etudes Politiques de Toulouse, 1983), pp. 220–221; Madeleine Rebérioux, "Histoire, historiens et dreyfusisme," *RH*, 518 (Apr.–June, 1976), 407–432. See also Benedetto Croce, "Les Etudes relatives à la théorie de l'histoire en Italie durant les quinze dernières années," *RSH*, 5 (1902), 257–269 and Karl Lamprecht, "La Méthode historique en Allemagne," *RSH*, 1 (1900), 21–27.

5 Steven Lukes, *Emile Durkheim, His Life and Work: A Historical and Critical Study* (Harmondsworth: Penguin, 1973), pp. 372–373.

6 Emile Durkheim, "Cours de science sociale, leçon d'ouverture," *RIE*, 15 (1888), 46–47. Cf. René Worms, "La Sociologie," *Revue Internationale de Sociologie*, 1:1 (1893), 3–16.

7 Emile Durkheim, "Les Règles de la méthode sociologique," *Revue Philosophique*, 37 (1894), 465–498; 38 (1894), 14–39, 168–182.

8 Paul Lacombe, *De l'histoire considérée comme science* (Paris: Hachette, 1894). In a later article Lacombe remarked that it mattered little to him whether one designated his form of history as history or sociology. Paul Lacombe, "L'Histoire comme science: à propos d'un article de M. Richert," *RSH*, 3 (1901), 9.

9 Cf. Gabriel Monod, "Bulletin historique: France, époque moderne," *RH*, 61 (1896), 325 and Henri Pirenne, "Une polémique historique en Allemagne," *RH*, 64 (1897), 56.

10 A.-D. Xénopol, *Les Principes fondamentaux de l'histoire* (Paris: Leroux, 1899), esp. chapters 1, 2, 8. For the debate between Lacombe and Xénopol see Paul Lacombe, "La Science de l'histoire d'après M. Xénopol," *RSH*, 1 (1900), 28–51; A.-D. Xénopol, "Les Faits de répétition et les faits de succession," *RSH*, 1 (1900), 121–136; A.-D. Xénopol, "Race et milieu," *RSH*, 1 (1900), 254–264; Paul Lacombe, "Milieu et race," *RSH*, 2 (1901), 34–55. For other related articles see Paul Lacombe, "La Méthode scientifique de l'histoire littéraire," *RSH*, 2 (1901), 153–166; Paul Lacombe, "L'Histoire comme science," pp. 2–9; A.-D. Xénopol, "La Classification des sciences et l'histoire," *RSH*, 2 (1901), 264–276; A.-D. Xénopol, "Les Sciences naturelles et l'histoire à propos d'un ouvrage récent," (Richert) *RSH*, 4 (1902), 276–292.

11 Charles V. Seignobos, *La Méthode historique appliquée aux sciences sociales* (Paris: Alcan, 1901); Terry N. Clark, *Prophets and Patrons: The French University and the Emergence of the Social Sciences* (Cambridge, MA: Harvard University Press 1973), pp. 156–159.

12 See, for example, Emile Durkheim, "Preface," *AS*, 1 (1898), iii; Victor Karady, ed., *Emile Durkheim: texts* (Paris: Editions de Minuit, 1975), I, pp. 199–200. Note also the discussion in Robert Bellah, "Durkheim and History," *American Sociological Review*, 24 (1959), 447–461.

13 See, for example, Seignobos, *La Méthode historique*, p. 161.

14 Ibid., p. 148.

15 Ibid., p. 153.

16 Ibid., p. 315.

17 Emile Durkheim, "Preface," p. iii.

18 Emile Durkheim, review of *La Méthode historique appliquée aux sciences sociales*, by Charles Seignobos, *AS*, 5 (1902), 123–127. Emile Durkheim, review of "La Storia considerata come scienza," by G. Salvemini and two other works, *AS*, 6 (1903), 125.

19 Emile Durkheim and Paul Fauconnet, "Sociologie et sciences sociales," *Revue Philosophique de la France et de l'Etranger*, 55 (1903), 485. See also Marcel Mauss and Paul Fauconnet, "Sociologie," *La Grande Encyclopédie*, xxx (Paris: Société Anonyme de la Grande Encyclopédie, 1901), p. 171.

20 Durkheim and Fauconnet, "Sociologie et sciences sociales," p. 493.

21 See discussion in Philippe Besnard, "The epistomological polemic: François Simiand," in Philippe Besnard, ed., *The Sociological Domain: The Durkheimians and the Founding of French Sociology* (Cambridge University Press and Paris: Editions de la MSH, 1983), pp. 248–249.

22 Berr had already published his own review of Seignobos, which had criticized Seignobos' empirical and highly restricted definition of social science, his lack of understanding of sociology, and his treatment of history as simply the history of events: Henri Berr, "Les Rapports de l'histoire et des sciences sociales d'après M. Seignobos," *RSH*, 4 (1902), 293–302. On the limited coverage given by other journals, see Rebérioux, "Le Débat," pp. 222–223.

23 François Simiand, "Méthode historique et science sociale, première partie," *RSH*, 6 (1903), 1–22.

24 "Société d'Histoire Moderne," *RIE*, 42 (1901), 242–243.

25 Ibid., "Deuxième partie," *RSH*, 6 (1903), 129–157. See also François Simiand, review of *Enseignement des sciences sociales*, by Henri Hauser, *NCSS*, 4 (Jan. 1903), 4–6. There Simiand attacked the use of separate monographs for each civilization.

26 François Simiand, "Méthode historique et science sociale, deuxième partie," *RSH* 6 (1903), 154–157.

27 François Simiand, "Questions à traiter et questions inutiles (histoire économique)," *NCSS*, 5 (Jan., 1904), 1–5.

28 On Simiand's address and the discussion which followed see "Séance du 3 janvier 1903," *BSHM*, 2 (1903), 73–77. See also "Séance du 7 février 1903," *BSHM*, 2 (1903), 79.

29 Paul Mantoux, "Histoire et sociologie," *RSH*, 7 (1903), 131.

30 Ibid., pp. 131–136. Cf. Lacombe's remarks in "Séance du 31 Mai, 1906: la causalité en histoire," *Bulletin de la Société Française de Philosophie* 6 (1906), 273.

31 Mantoux, "Histoire et sociologie," pp. 131–134.

32 François Simiand, "Anthropomorphisme et finalisme," *NCSS*, 5, 33 (Mar. 1904), 74.

33 Clark, *Prophets,* pp. 159–160.

34 For reports of Seignobos' conference see E. F., "Ecole des Hautes Etudes Sociales: les rapports de la sociologie avec l'histoire," *RU*, 13: 1 (1904), 21–24 and Marcel Pournin, "La Sociologie et les sciences sociales: rapports de la sociologie avec l'histoire," *Revue International de Sociologie*, 12 (Mar. 1904), 161–165. Cf. the addresses of Bouglé and Lanson who argued that history required sociological explanations. Pournin, "La Sociologie," pp. 166–167; E. F., "Ecole des Hautes Etudes Sociales: les rapports de la sociologie avec l'histoire de la littérature," *RU*, 13, 1 (1904), 233.

35 "Séance du 31 Mai 1906," pp. 274–277.

36 "Séance du 30 Mai 1907: les conditions pratiques de la recherche des causes dans le travail historique," *Bulletin de la Société Française de Philosophie*, 7: 7 (July 1907), 283.

37 Ibid., p. 273.

38 Ibid., pp. 290, 304, 299. See also comments of Glotz, in ibid., p. 303.

39 Ibid., pp. 293, 285, 306, 308–309.

40 "Séance du 28 Mai 1908: l'inconnu et l'inconscient en histoire," *Bulletin de la Société Française de Philosophie*, 8 (1908), 225.

41 Discussion for meeting of May 28, 1908 reproduced in Karady, ed., *Emile Durkheim*, pp. 199–217. This statement was not unlike that of Simiand in 1907 who had opened his comments with the observation, "Je crois bien, pour ma part, que finalement l'effort scientifique en histoire aboutira à une sociologie." "Séance du 30 mai 1907," p. 292.

42 Karady, ed., *Emile Durkheim*, pp. 203, 201.

43 Ibid., p. 202. Cf. Comments of Lacombe, who remained much more open to Durkheim's arguments (p. 213).

44 Ibid., p. 211. On the contrast between Gustave Bloch and Durkheim see also: Emile Durkheim, review of *La Plèbe romaine*, by Gustave Bloch, *AS*, 12 (1913), 441–443. There Durkheim criticized Gustave for depicting *la plèbe* and *le patriciat* as specifically Roman, and argued that comparisons should have been drawn to Athens. Of the ban on "connubium" Durkheim wrote, "Ce n'est pas un fait purement 'romain' comme dit notre auteur, on le retrouve partout où des classes fortement constituées s'opposent l'une à l'autre." Whereas Durkheim sought comparisons and laws, Gustave stressed the particular character of the Roman society which he studied.

45 On Bouglé's association with the other Durkheimians, see W. Paul Vogt, "Durkheimian sociology versus philosophical rationalism: the case of Célestin Bouglé," in Besnard, ed., *Sociological Domain*, pp. 231–247.

46 Karady, ed., *Emile Durkheim*, pp. 211–212.

47 On Durkheim's avoidance here, see Rebérioux, "Le Débat," p. 227.

48 Karady, ed., *Emile Durkheim*, p. 217.

49 Marc Bloch, "Méthodologie historique," pp. 155–159.

50 Ibid., pp. 159–161.

51 Ibid., p. 162.

52 Ibid., pp. 161, 159.

4 The quest for identity in Vidalian geography

1 P. Vidal de la Blache, "La Géographie humaine, ses rapports avec la géographie de la vie," *RSH*, 7 (1903), 240.

2 Although like Durkheim both a Dreyfusard and a member of the Union pour la Verité, Vidal's direct political involvement was mostly limited to discussions of regionalism and boundary disputes both within France and between France and other powers. For Vidal's political involvement, see Georges Nicolas, "Vidal de la Blache et la politique," *Bulletin de l'Association de Géographes Français*, 65: 4 (1988), 333–337; André-Louis Sanguin, "Vidal de la Blache et la géographie politique," *Bulletin de l'Association de Géographes Français*, 65: 4 (1988), 321–331. On the political implication of Vidal's *Tableau de la géographie de la France,* see Marie-Claire Robic, "National Identity in Vidal's *Tableau de la géographie de la France:* From political geography to human geography," in David Hooson, ed., *Geography and National Identity* (Oxford: Blackwell, 1994), pp. 58–70.

3 The complex issue of the differences between Ratzel and Vidal is touched upon but not exhausted in a number of publications. Vidal was less ambitious and systematic than Ratzel; he tempered some of the more deterministic elements of Ratzel's geography but certainly was not totally immune from them. See, for example, Numa Broc, "La Géographie française face à la science allemande, 1870–1914," *AG,* 86 (1977), 88–89; Anne Buttimer, *Society and Milieu in the French Geographic Tradition* (Chicago: Rand McNally, 1971), AAG monograph no. 6, pp. 44–46; Jean-Yves Guiomar, "Le *Tableau de la Géographie de la France* de Vidal de la Blache," pp. 595, note 5 in Pierre Nora, ed., *Les Lieux de mémoire,* II, *La Nation,* vol. 1 (Paris: Gallimard, 1986); André-Louis Sanguin, *Vidal de la Blache: un génie de la géographie* (Paris: Belin, 1993), p. 136.

On the emergence of the term *géographie humaine,* see Marie-Claire Robic, "L'Invention de la 'géographie humaine' au tournant des années 1900: les Vidaliens et l'écologie," in Paul Claval, ed., *Autour de Vidal de la Blache: la formation de l'école française de géographie* (Paris: CNRS, 1993), pp. 137–147.

4 Clearly, the methodological statements made by Vidal de la Blache and his students at the turn of the century were not solely in response to criticism by the Durkheimians. As they attempted to professionalize geography and turn it into a distinct scientific field they also needed to consider other potential challengers, including within France the followers of Le Play (a Catholic social reform movement), geography as it then existed in France in both its scholastic and historical forms, geology, and German geography. However, for the sake of clarity, I will focus here simply on their positions *vis-à-vis* the Durkheimians.

5 P. Vidal de la Blache and Marcel Dubois, "Avis au lecteur," *AG,* 1 (1891), i–iv.

6 "Le Principe de la géographie générale," *AG,* 5 (1895–1896), pp. 129–142.

7 P. Vidal de la Blache, "Des divisions fondamentales du sol français," *Bulletin Littéraire,* 2: 1 (Oct. 10, 1888), 1–2; P. Vidal de la Blache, "La Géographie politique," *AG,* 7 (1898), 97–111.

8 Vidal de la Blache, "La Géographie politique," p. 111. P. Vidal de la Blache, "Leçon d'ouverture du cours de géographie," *AG,* 8 (1899), 97–109.

9 (Emile Durkheim), unsigned review of *Der Staat und sein Boden geographisch beobachtet,* by F. Ratzel, *AS,* 1 (1898), 538. In a later review Durkheim also argued that contrary to Ratzel's claims, societies became less dependant on their material bases as they developed. See Emile Durkheim, review of *Das Meer als Quelle der Völkergrösse,* by F. Ratzel, *AS,* 4 (1900), 567.

10 Emile Durkheim, "Preface," *AS,* 2 (1899), iii–iv.

11 Emile Durkheim, "Morphologie sociale," *AS*, 2 (1899), 521.
12 Emile Durkheim, review of *Politische Geographie*, by F. Ratzel, *AS*, 2 (1899), 531–532.
13 Emile Durkheim, review of *Anthropogeographie*, I, by F. Ratzel, *AS*, 3 (1900), 557. In this review, Durkheim used the phrase "preponderant cause" in much the same way as he had used "essential cause" the year before.
14 See Durkheim, review of *Politische Geographie*, p. 531.
15 Vidal de la Blache, "La Géographie politique," p. 99.
16 (Emile Durkheim), unsigned note re "La Géographie politique," by P. Vidal de la Blache, *AS*, 2 (1899), 532.
17 Emile Durkheim, review of *Les Grandes Routes des peuples*, by E. Demolins, *AS*, 5 (1902), 560–562; Emile Durkheim, review of same, *NCSS*, 2: 5 (1901), 152–153; Emile Durkheim, review of two articles by F. Schrader, *AS*, 6 (1903), 539–540.
18 M. Mauss and P. Fauconnet, "Sociologie," in *La Grande Encyclopédie*, xxx (Paris: Société Anonyme de la Grande Encyclopédie, 1901), p. 175.
19 F. Simiand, "'Géographie humaine' et sociologie," *NCSS*, 4 (1903), 51. For Simiand's reactions to Henri Hauser's defence of geography, see F. Simiand, "Méthode historique et science sociale," *RSH*, 6 (1903), 132; F. Simiand, "Sur le plan des *Notes Critiques*," *NCSS*, 4 (1903), 3–4.
20 Marcel Mauss and Henri Beuchat, "Essai sur les variations saisonnières des sociétés Eskimos: étude de morphologie sociale," *AS*, 9 (1906), 39–132. On Beuchat's limited contribution to this work, see Marcel Mauss, "An intellectual self-portrait," in Philippe Besnard, ed., *The Sociological Domain: The Durkheimians and the Founding of French Sociology* (Cambridge University Press and Paris: Editions de la MSH, 1983), p. 141.
21 Mauss and Beuchat, "Essai," p. 127.
22 Mauss and Fauconnet, "Sociologie," p. 172. Cf. Georges Davy, "La Sociologie de M. Durkheim," *Revue Philosophique*, 72 (1911), p. 167.
23 Mauss and Fauconnet, "Sociologie," p. 172. Mauss returned to this argument in a 1905 review and in 1911 Davy would also indicate that the social substratum and the collective life were intimately related. Marcel Mauss, review of *Völkerkunde*, by H. Schurtz (and two other works), *AS*, 8 (1905), 166–167; Davy, "La Sociologie de M. Durkheim," 52, 167ff.
24 Emile Durkheim, "Sociologie et sciences sociales," in H. Bouasse, et al., *De la méthode dans les sciences* (Paris: Alcan, 1909), p. 272.
25 Antoine Vacher, review of *La Valachie*, by E. de Martonne, *NCSS*, 4 (1903), 56.
26 P. Vidal de la Blache, "Les Conditions géographiques des faits sociaux," *AG*, 9 (1902), 23.
27 Vidal de la Blache, "La Géographie humaine, ses rapports," pp. 220–222.
28 Robic, "L'Invention," p. 142. On the contribution of other "Vidalians" to the "biological conception" of geography see the interesting essay, Paul Claval, "Le Rôle de Demangeon, de Brunhes et de Gallois dans la formation de l'école française: les années 1905–1910," in Claval, ed., *Autour de Vidal de la Blache*, pp. 149–158. See also the interesting argument by Berdoulay that the naturalistic language adopted by Vidal and his followers was ill suited to advancing "possibilism" and as such may well have contributed to its decline, Vincent Berdoulay, "La Géographie Vidalienne: entre texte et contexte," in Claval, ed., *Autour de Vidal de la Blache*, p. 26.

29 Vidal de la Blache, "La Géographie humaine, ses rapports," p. 232.
30 Ibid., pp. 223–224 (author's emphasis).
31 Ibid., p. 240.
32 P. Vidal de la Blache, "Rapports de la sociologie avec la géographie," *RIS*, 12: 5 (May 1904), 312.
33 Ibid., p. 313.
34 Vidal de la Blache, "La Conception actuelle de l'enseignement de la géographie," *AG*, 14: 74 (1905), 193–207.
35 Victor Duruy, *Histoire de France* (Paris: Hachette, 1860); Jules Michelet, *Tableau de la France* (Paris: Société des Belles Lettres, 1949) (originally, 1833).
36 P. Vidal de la Blache, *Tableau de la géographie de la France* (Paris: Hachette, 1903), I, part 1 of Ernest Lavisse, *Histoire de France depuis les origines jusqu'à la Révolution.* For an interesting comparison of Michelet's work and Vidal's see Guiomar, "Le *Tableau*," pp. 574–576.
37 Jean-Claude Bonnefort, "La Lorraine dans l'oeuvre de Paul Vidal de la Blache," in Claval, ed., *Autour de Vidal de la Blache,* pp. 84–85.
38 On Vidal's mandate to present *une France d'autrefois,* see Daniel Loi, Marie-Claire Robic, and Jean-Louis Tissier, "Les Carnets de Vidal de la Blache, esquisses du *Tableau?*" *Bulletin de l'Association de Géographes Français,* 65: 4 (1988), 307–308.
39 Vidal's interest in the local life was demonstrated in his notebooks which contained many (albeit somewhat distant) observations on it, but he included little on that life in the *Tableau.* Ibid., p. 307.
40 See Vidal de la Blache, "Régions françaises," *Revue de Paris* (Dec. 15, 1910), 821–849; Vincent Berdoulay, *La Formation de l'école française de géographie (1870–1914)* (Paris: Bibliothèque Nationale, 1981), Mémoires de la Section de Géographie no. 11, p. 135; Guiomar, "Le *Tableau*," p. 575.
41 Bertrand Auerbach, review of *Tableau*, by Vidal de la Blache, *Revue Générale des Sciences Pure et Appliqués,* 14 (1903), 897. Note also E. Levasseur, review of *Tableau*, by Vidal de la Blache, *Journal des Savants,* 1 n.s. (1903), 617; Emile Chantriot, review of *Tableau*, by Vidal de la Blache, *RSH*, 7 (1903), 59. Gallois in his review commented that Vidal had managed to undo some of the damage to the reputation of *géographie humaine* caused by Michelet's "tendance dangereuse d'essayer de caractériser le génie d'une province par celui des écrivains qu'elle a produit." Lucien Gallois, review of *Tableau*, by Vidal de la Blache, *AG*, 12 (1903), 208.
42 Gabriel Monod, "Rapport sur le concours pour le Prix Audiffred," *ASMP Compte-Rendu,* 66 n.s. (1906), 335. See also, Georges Weuleresse, review of *Tableau*, by Vidal de la Blache, *Revue Pédagogique,* 44 (1904), 305; Ch. Dufayard, review of *Tableau*, by Vidal de la Blache, *RU*, 12:2 (1903), 244–245. For a recent assessment see Guiomar, "Le Tableau," pp. 569–597. In 1906, reflecting his enhanced stature, Vidal was elected to the Académie to fill the chair left vacant by the death of Albert Sorel. See Sanguin, *Vidal de la Blache,* pp. 200–201.
43 See for example, DP Antoine Vacher to Albert Demangeon, Feb. 8, 1905.
44 Antoine Vacher, review of *Tableau*, by Vidal de la Blache, *AS*, 8 (1905), 614–615.
45 The accepted view has been that the *maître* Vidal had a profound and inspiring effect on his students. See for example, André Meynier, *Histoire de la pensée géographique en France* (Paris: PUF, 1969), p. 30. The testimony there said to be by

Demangeon was actually that of de Martonne. See Maurice Grandazzi, "Une hommage à la mémoire d'Albert Demangeon," *AG*, 51 (1942), 304.

46 DP Antoine Vacher to Albert Demangeon, Jan. 9, 1909.

47 See, for example, DP letters of Antoine Vacher to Albert Demangeon, Dec. 11, 1908; Jan. 9, 1909; Jules Sion to Albert Demangeon, n.d. 1909; Friday 13, 1910.

48 The six monographs all written by former students of the ENS under Vidal's supervision were, in order of publication: Albert Demangeon, *La Picardie et les régions voisines* (Paris: Colin, 1905); Raoul Blanchard, *La Flandre* (Paris: Colin, 1906); Raoul de Félice, *La Basse-Normandie* (Paris: Coueslant, 1907), also published by Hachette; Camille Vallaux, *La Basse-Bretagne* (Paris: Colin, 1908); Antoine Vacher, *Le Berry* (Paris: Colin, 1908); Jules Sion, *Les Paysans de la Normandie Orientale* (Paris: Colin, 1909).

49 See interesting discussion in Roger Chartier, "Science sociale et découpage régional: note sur deux débats, 1820–1920," *Actes de la Recherche en Sciences Sociales*, 35 (1980), 27–36.

50 See Lucien Gallois, *Régions naturelles et noms de pays* (Paris: Colin, 1908), p. 224; P. Vidal de la Blache, review of *Régions naturelles et noms de pays*, by L. Gallois, *Journal des Savants* (Oct. 1909), 456–457.

51 Demangeon, *La Picardie*, pp. 7, 420; de Félice, *La Basse-Normandie*, p. 7.

52 Linked to both Social Darwinism and Lamarckian environmentalism, environmental determinism involved the causal explanation of human phenomena by the physical environment. It marked a considerable body of geographical works in the early twentieth century. In the United States, where key proponents included Ellen Churchill Semple and to a lesser degree Ellsworth Huntington, it came to mar the reputation of geography as a field. French geographers were not immune from this line of argument as can be seen in the works of Edmond Demolins and some of Vidal's early works but were not as quick to adopt it. See Jurgen Herbst, "Social Darwinism and the history of American geography," *Proceedings of the American Philosophical Society*, 105:6 (1961), 538–533; David Livingstone, *The Geographical Tradition: Episodes in the History of a Contested Enterprise* (Oxford: Blackwell, 1992), pp. 177–259.

53 Vacher, *Le Berry*, p. 60.

54 See, for example, Sion, *Les Paysans*, pp. 139, 186, 259, 311.

55 Ibid., pp. 110–111, 409.

56 Febvre's reviews of these monographs were as follows: Lucien Febvre, "Une région géographique: La Flandre," *RSH*, 14 (1907), 92–94 (review of Blanchard); Lucien Febvre, "Une étude de géographie humaine," *RSH*, 16 (1908), 45–49 (review of Vallaux); Lucien Febvre, "A propos d'une monographie géographique," *RSH*, 17 (1908), 358–360 (review of de Félice); Lucien Febvre, "Une étude géographique sur le paysan normand," *RSH*, 19 (1909), 43–51 (review of Sion); Lucien Febvre, "Sur quelques ouvrages de géographie et d'histoire régionales," *RSH*, 19 (1909), 102–104 (review of Vacher).

57 AN F^{17} 25319.

58 AN 360 AP 2 (Fonds Henri Wallon), letter from Sion and Febvre to Henri Wallon, Mar. 23, 1912; postcard from Febvre and Sion to Wallon postmarked 1913. See also letter from Sion to Wallon, Aug. 8, 1905.

59 De Félice entered the ENS in 1889, Blanchard in 1897, and both Vacher and

Demangeon worked there as *maîtres surveillants* at the turn of the century. Furthermore, Demangeon was the brother-in-law of Henri Wallon, another of Febvre's closest friends from the *promotion* of 1899. Re Febvre, Wallon, and Demangeon, see AN 360 AP 2 (Fonds Henri Wallon), letters to Febvre and Wallon from Jules Bloch, July 8, 1903 and from Jules Sion, Dec. 31, 1907; DP letter Febvre to Demangeon, Sat. 30, 1905.

60 Lucien Febvre, *Combats pour l'histoire* (Paris: Colin, 1953), p. 381. See also Febvre's very favorable comments re Demangeon in Febvre, "Une région géographique: La Flandre," p. 92 and in Febvre, "Une étude géographique sur le paysan normand," p. 43.

61 Febvre, "A propos d'une monographie géographique," pp. 359–360.

62 This phrase refers to the tendency of some regional geographies to treat each topic separately, placing the various subfields of physical and human geography in individual "drawers."

63 Febvre, "Sur quelques ouvrages," p. 102. See also discussion above re DP Antoine Vacher to Albert Demangeon, Jan. 9, 1909, on pp. 66–67 of this chapter.

64 Febvre, "Une étude géographique sur le paysan normand," p. 51. Cf. P. Vidal de la Blache, review of *Les Paysans de la Nomandie Orientale*, by Jules Sion, *AG*, 18 (1909), 178.

65 DP Jules Sion to Albert Demangeon, Feb. 8, 1909.

66 See the interesting discussion in Chartier, "Science sociale et découpage régional," pp. 32–36.

67 François Simiand, review of *La Picardie* and others, by Albert Demangeon and others, *AS*, 11 (1910), 731.

68 P. Vidal de la Blache, "Les Genres de vie dans la géographie humaine," *AG*, 20 (1911), 304.

69 See, for example, discussions in Marie-Claire Robic, "La Conception de la géographie humaine chez Vidal de la Blache d'après les *Principes de géographie humaine*," *Les Cahiers de Fontenay*, 4 (1976), 19–20; Guiomar, *Le Tableau*, pp. 581–582; Jean-Marc Besse, "Idéologie pour une géographie: Vidal de la Blache," *Espaces-Temps*, 7:12 (1979), 75, 78–87; Robic, "National Identity," p. 65. In this last article Robic notes the contrast between Vidal's use of ethnographic material and that of the folklorists who paid much more attention to dress, a subject which did not fit well into Vidal's "naturalist conceptions of the relationship between man, as a biological species, and his environment."

70 P. Vidal de la Blache, "Sur le sens et l'objet de la géographie humaine," parts 1 & 2, *Revue Bleue*, 50 (Apr. 27, 1912), 513.

71 Cf. Robic, "La Conception," pp. 1–76. As Claval and others have noted, in Vidal's later substantive work (particularly that on regional definition), he became increasingly open to the study of urban, industrial and even political factors. Nevertheless, he would continue to draw on biological metaphors. See, for example, Paul Vidal de la Blache, "La Relativité des divisions régionales," in Camille Bloch, et al., *Les Divisions Régionales de la France* (Paris: Alcan, 1913), p. 13; Paul Vidal de la Blache, "Régions françaises," *La Revue de Paris*, 6 (1910), 832, 849. Cf. Paul Claval, "Le Rôle de Demangeon, de Brunhes et de Gallois," p. 149.

This use of biological metaphors was, however, less true of his *La France de l'Est*, which was addressed to a more general public and not shaped as a specifically geographical work. See Paul Vidal de la Blache, *La France de l'Est* (Paris: Colin, 1917), pp. 1–7; Yves Lacoste, "A bas Vidal. . . Viva Vidal," *Hérodote*, 16 (1979), 72–77.

72 P. Vidal de la Blache, "Sur le sens et l'objet de la géographie humaine," parts 3 & 4, *Revue Bleue*, 50 (May 4, 1912), 548–549.

73 P. Vidal de la Blache, "Des caractères distinctifs de la géographie," *AG*, 22 (1913), 289–299.

74 Ibid., p. 299.

75 P. Vidal de la Blache, "Sur l'esprit géographique," *Revue Bleue* (1914), 558.

76 Vidal de la Blache, "Des caractères distinctifs," p. 289.

77 Vidal de la Blache, "Sur l'esprit géographique," p. 560.

78 Emile Durkheim, review of *La Géographie humaine*, by Jean Bruhnes, *AS*, 12 (1913), 819–821.

79 Albert Demangeon, review of *Influences of Geographic Environment*, by Ellen Churchill Semple, *AS*, 12 (1913), 811–812.

80 Albert Demangeon, review of *Géographie sociale: le sol et l'état*, by Camille Vallaux, *AS*, 12 (1913), 814–818. Durkheim also approached Demangeon to review *La Géographie humaine* by Brunhes. See DP letter of Emile Durkheim to Albert Demangeon, Sept. 20, 1912.

81 There is even less to go on regarding Bloch's reactions to Ratzel. Not only was he exposed to discussions of Ratzel by both Vidalians and Durkheimians, he also probably had more direct contact with German geography during 1908 when he studied at Leipzig. However, apart from the very occasional use of the term "anthropogeography," direct references to German geography would be limited to just a few post-war reviews. Cf. Pierre Toubert, "Preface," to Marc Bloch, *Les Caractères originaux de l'histoire rurale française* (Paris: Colin, 1988), pp. 89.

5 From the Fondation Thiers to the doctorate: Marc Bloch's emerging perspective

1 "La Fondation Thiers," *Annuaire de la Fondation Thiers*, 1947 (for the years 1941–1946), inside front cover.

2 Emile Boutroux, "Rapport de M. le Directeur au Conseil Administratif (Oct. 10, 1912)," *Annuaire de la Fondation Thiers*, n.s. 11 (1913), 6.

3 As the librarian at the ENS, Lucien Herr promoted a moderate form of socialism associated with Jean Jaurès. He attracted a very active following among the *normaliens* including many of the young Durkheimians. See Philippe Besnard, ed., *The Sociological Domain: The Durkheimians and the Founding of French Sociology* (Cambridge University Press and Paris: Editions de la M.S.H., 1983), pp. 24–25. Both wrote "Cahiers" for this group. According to Yves Goudineau, Granet's "Contre l'alcoolisme" was inspired by Durkheim's *Suicide*. Yves Goudineau, "Introduction à la sociologie de Marcel Granet," thèse, U. de Paris X, Laboratoire d'ethnologie et de sociologie comparée, p. 32.

4 See Emile Boutroux, "Rapport de M. le Directeur" for the years 1908–1912, *Annuaire de la Fondation Thiers* (1909–1913). Direct influences of the Nouvelle Sorbonne were less evident in Leroux's study of Anglo-American pragmatism and in Loubers' legal studies.

5 See Emile Boutroux, "Rapport de M. le Directeur" for the years 1911–1914, *Annuaire de la Fondation Thiers* (1912–15).

6 For a list of all the *pensionnaires* up to 1914 see *Annuaire de la Fondation Thiers*, n.s. 11 (1913), 16–17.

7 The evidence here calls into question the Bryce Lyon's assertion, "World War I led him [Bloch] to concentrate his research upon ordinary people and to use social sciences to gain information and insights on their behavior." These interests were already strongly developed during his tenure at the Fondation Thiers. Bryce Lyon, "Introduction," to Lucien Febvre and Marc Bloch, *The Birth of Annales History: The Letters of Lucien Febvre and Marc Bloch to Henri Pirenne (1921–1935)*, eds. Bryce Lyon and Mary Lyon (Brussels: Académie Royale de Belgique, Commission Royale d'Histoire, 1991), p. xxxv.

8 AN 318 MI 1, no. 433–436, letter of Marc Bloch to Emile Boutroux, May 1, 1908.

9 Emile Boutroux, "Rapport de M. le Directeur," *Annuaire de la Fondation Thiers* (1910), 13–14; (1911), 11–12; (1912), 8.

10 See discussion in Jacques Le Goff's preface to Marc Bloch, *Les Rois thaumaturges* (Paris: Gallimard, 1983), pp. iv–v.

11 Georges Davy, "Louis Gernet, l'homme et le sociologue," in Marcel Bataillon et al., *Hommage à Louis Gernet* (Paris: PUF, 1966), p. 8.

12 Henri Lévy-Bruhl, "Louis Gernet, historien du droit grec," in Bataillon et al., *Hommage*, p. 15.

13 See AN 318 MI 1, no. 487–512, letters of Marc Bloch and Gustave Bloch to Georges Davy, c. 1908–1924. In fact, Bloch's relationships with all of these colleagues did not begin at the Fondation Thiers as the tenures of all three at the ENS had also overlapped Bloch's (Gernet entered in 1902, Granet in 1904, and Davy in 1905). Furthermore, both Bloch and Granet had both attended the Lycée Louis-le-Grand.

14 Louis Gernet, "L'Approvisionnement d'Athènes en blé au Ve et au VIe siècles," in Gustave Bloch, *Mélanges d'histoire ancienne* (Paris: Alcan, 1909), pp. 269–391. (Bibliothèque de la Faculté des Lettres de Paris, 25); François Simiand, review of above, *AS*, 11 (1910), 563.

15 For early research interests of Marcel Granet and Georges Davy see Emile Boutroux, "Rapport de M. le Directeur," for the years 1909–1912 in the *Annuaire de la Fondation Thiers* (1910–1913).

16 Henri Berr, "Les Régions de la France," *RSH*, 6 (1903), 180.

17 It does appear, however, that Demangeon initially agreed to contribute to Berr's series. See DP: letter of Henri Berr to Albert Demangeon, Aug. 30, 1905.

18 The occupations of these men were taken from the volumes of *Minerva* for these years. No occupation was found for Louis Villat. At the time, Pfister was a professor of medieval history at the Sorbonne, but he too had previously held a chair of regional history at the Faculty of Letters of Nancy.

19 Some became more directly involved in the debates over *sociétés savantes*. Barrau-Dihigo endorsed Dumoulin's proposal for a federation of the societies and Kleinclausz echoed earlier remarks by Pierre Caron, suggesting that the universities take an active role in leading the societies and coordinating their activities. L. Barrau-Dihigo, "La Gascogne," *RSH*, 6 (1903), 298; A. Kleinclausz, "La Bourgogne," *RSH*, 9 (1904), 195–197.

20 Marc Bloch, "L'Ile-de-France, III & IV," *RSH*, 26 (1913), 151.

21 Ibid., p. 348.

22 Barrau-Dihigo, for example, argued that *gascons* were *fier, vantard, résolu, prudent, hardi, vif*, and so forth. Barrau-Dihigo, "La Gascogne," pp. 282–290.

23 Sebastien Charléty, "Le Lyonnais," *RSH*, 8 (1904), 45.

24 For Bloch's reaction to the type of analysis which characterized many of the monographs, see his later review: Marc Bloch, review of *La Vie agricole dans la Picardie Orientale depuis la guerre*, by A. Arsène Alexandre, *AHES*, 1 (1929), 102.

25 The one exception was Villat. See Louis Villat, "Le Velay," *RSH*, 16 (1906), 308–310.

26 Henri Prentout, "La Normandie, I & II," *RSH*, 19 (1909), 52.

27 For example, Barrau-Dihigo noted that "Gascogne" despite its lack of precise and stable boundaries had always revolved around Armagnac; Charléty said of Lyonnais, "la tête seule importe"; and Villat described Le Puy as the center of attraction for Le Velay. Barrau-Dihigo, "La Gascogne," p. 184; Charléty, "Le Lyonnais," p. 45; Villat, "Le Vilay," pp. 303–310.

28 Cf. Lucien Gallois, *Régions naturelles et noms de pays* (Paris: Colin, 1908).

29 Marc Bloch, "L'Ile-de-France," 1913, p. 349.

30 In Pfister's case the section on *pays* was actually written by his friend Gallois. See Christian Pfister, "La Lorraine, le Barrois et les Trois-Evêchés," *RSH*, 22 (1911), 164, note 1. Febvre, though arguing, "Ce n'est pas un être géographique que la Franche-Comté," gave a detailed description of the internal physical geography stressing the geology, soils, and climate. Lucien Febvre, "La Franche-Comté," *RSH*, 10 (1905), 187–193.

31 Marc Bloch, "L'Ile-de-France," 1913, p. 350.

32 Ibid., 1913, p. 147.

33 Ibid., pp. 148, 152.

34 Marc Bloch, "L'Ile-de-France, I & II," *RSH*, 25 (1912), 339. Cf. Marc Bloch, "Ile-de-France," reprinted in Marc Bloch, *Mélanges historiques* (Paris: Serge Fleury and EHESS, 1983), II, p. 724.

35 Marc Bloch, "L'Ile-de-France," in Marc Bloch, *Mélanges historiques*, II, p. 700.

36 Marc Bloch, "L'Ile-de-France," 1913, p. 154. In the English translation *l'histoire géographique* was mistakenly translated as "historical geography." Marc Bloch, *The Ile-de-France: The Country around Paris* (Ithaca: Cornell University Press, 1971), p. 64. (J. E. Anderson, trans.)

37 Marc Bloch, "L'Ile-de-France," 1913, p. 147.

38 See the introductory essay by Carole Fink to Marc Bloch, *Memoirs of War, 1914–1915* (Ithaca: Cornell University Press, 1980), p. 25.

39 Marc Bloch, review of *Formation et évolution de Paris*, by Marcel Poète, *RSH*, 23 (1911), 378–379; Marc Bloch, review of *Origines et formation de la nationalité française. Elements ethniques, unité territoriale*, by Auguste Longnon, *RSH*, 25 (1912), 365. See also Bloch's notes on Longnon's works in AN AB[xix] 3796.

40 Febvre's thesis on Franche-Comté was reviewed for Berr by Prentout, an established historian with an interest in the Reformation. See Henri Prentout, "Philippe II et la Franche-Comté," *RSH*, 25 (1912), 59–65.

41 Lucien Febvre, "Marc Bloch et Strasbourg: souvenirs d'une grande histoire," in *Mémorial des années 1939–1945* (Paris: Les Belles Lettres, 1947), Publications de la Fac. des Lettres de l'U. de Strasbourg, fasc. 103, pp. 171–2.

42 Marc Bloch, review of *Histoire de Franche-Comté*, by Lucien Febvre, *RSH*, 28 (1914), 354–355.

43 Ibid., p. 356. Bloch would remain a staunch opponent of such characterizations of regional "mentalités." See, Marc Bloch, review of *La Vie agricole dans la Picardie Orientale*, p. 102.

44 See AN 318 MI 1, no. 439, letter of reference by Gallois, n.d. (testifying to Bloch's interest in geography); EB Dossiers: "Géographie Générale"; "Les Trois Péninsules." The former contains notes on Lapparent, Schirmer, and de Martonne, as well as on a course, "Leçons de géographie générale"; the latter appears to be a regional geography course, though whether this is a course Bloch taught or one he took is not clear.

45 According to Perrin, Bloch as a veteran was allowed to forgo the usual requirement for two theses, a principal and a complementary thesis. Ch.-Edmund Perrin, "L'Oeuvre historique de Marc Bloch," *RH*, 199 (1948), 163. However, Bloch's files indicate that he submitted and defended his article: Marc Bloch, "Les Formes de la rupture de l'hommage dans l'ancien droit féodal," *Nouvelle Revue Historique du Droit Français de Etranger*, 36 (Mar.–Apr. 1912), 141–177. See AN AB^{XIX} 3811. The official register lists this article as his principal thesis and *Rois et serfs* as his complementary one, no doubt an error, AN AJ* 4764.

46 Marc Bloch, *Rois et serfs: un chapitre d'histoire Capétienne* (Geneva: Slatkine-Megariotis, 1976), p. 132. In translation this statement would read, "In general, throughout our kingdom, as much as it concerns us and our successors, may such servitudes be brought to an end."

47 Ibid., p. 131.

48 Nevertheless, in one case Bloch did note fundamental differences in the character of serfdom. Heavier financial burdens on serfs in Languedoc led to particularly strong opposition to their collection. Marc Bloch, *Rois et serfs*, p. 104.

49 While at the Fondation, Gernet, for example, had published an article on the ancient Greek word for murderer which advocated a careful examination of word usage in surviving texts. Gernet noted changing and ambiguous meanings and intense usages and examined the grammar to trace and uncover fundamental "collective representations." By contrast, Bloch simply argued through an examination of enduring images. Louis Gernet, "Authentes," *Revue des Etudes Grècques*, 22 (1909), 13–32.

50 Cf. Gernet, "L'Approvisionnement d'Athènes en blé."

51 Marc Bloch, *Rois et serfs*, p. 176.

52 Ibid., p. 28.

53 Ibid., pp. 41–42.

54 Ibid., pp. 45–46.

55 Georges Davy, *La Foi jurée: étude semantique du problème du contrat* (Paris: Alcan, 1922), pp. 17, 75, 112, 125, 248, 289, 313, 373; Marcel Granet, *Fêtes et chansons anciennes de la Chine* (Paris: Leroux, 1919) – Granet's principal thesis; Marcel Granet, "Coutumes matrimoniales de la Chine Antique," *Toung-pao*, 13 (1912), 63–94 – an earlier version of the argument in his thesis.

56 Marc Bloch, *Rois et serfs*, p. 161. Cf. Louis Gernet, *Recherches sur le développement de la pensée juridique et morale en Grèce; étude semantique* (Paris: Leroux, 1917); Louis Gernet, "Hypothèses sur le contrat primitif en Grèce," *Revue des Etudes*

Grècques, 30 (1917), 249–293, 363–383; Gernet, "Authentes"; Davy, *La Foi jurée* (see, for example, pp. 74–75, 371–374).

57 Marc Bloch, *Rois et serfs*, p. 12.

58 Ibid., p. 13.

59 This phrase is taken from Georges Davy, "Phénomènes collectifs et institutions sociales," *Revue de Metaphysique et Morale*, 26 (1919), reprinted in Georges Davy, *L'Homme: le fait social et le fait politique* (Paris: Mouton, 1973), p. 43.

60 Marc Bloch, *Rois et serfs*, p. 54.

61 Cf. ibid., p. 166.

62 Marc Bloch, "Les Formes de la rupture de l'hommage," as reprinted in Bloch, *Mélanges historiques*, 1, pp. 189–190.

63 Ibid., p. 209. Similarly, Bloch felt that an interesting study could be done comparing the forms of "défi" to those of the breaking of homage.

64 An interesting contrast can be drawn between Bloch's rather skeptical approach and that of Davy in his review of a work by von Amira which dealt with the symbolism of the stick or staff in ancient German law. Here, Davy seemed much less critical than Bloch of sweeping characterizations of the meaning and origins of the symbol of the stick or staff. Georges Davy, review of *Der Stab in der Germanischen Rechtssymbolik*, by K. von Amira, *AS*, 11 (1913), 544.

65 Marc Bloch, review of *An Introduction to the Study of Prices*, by Walter Layton, *RSH*, 25 (1912), 106; Marc Bloch, "Note sur deux ouvrages d'histoire et d'économie rurales," *RSH*, 27 (1913), 165, 167.

66 Marc Bloch, review of *Der Deutsche Staat des Mittelalters*, by G. von Below, *RH*, 128 (1918), 345. See also, Marc Bloch, review of *Geldwert in der Geschichte. Ein methodologischer Versuch*, by Andreas Walther, *RSH*, 25 (1912), 244; Marc Bloch, review of *An Introduction to the Study of Prices*, p. 106.

67 Charles-Victor Langlois and Charles Seignobos, *Introduction aux études historiques* (Paris: Hachette, 1898).

68 Marc Bloch, "Critique historique et critique du témoignage," *AESC*, 5 (1950), 8. (Originally given as a lycée address on July 13, 1914.)

69 Marc Bloch, "Notes sur les sources de l'histoire de l'Ile-de-France au Moyen Age," *Bulletin de la Société d'Histoire de Paris et de l'Ile-de-France*, 40 (1913), 161. See also, Marc Bloch, "Blanche de Castille et les serfs du Chapitre de Paris," *Mémoires de la Société de l'Histoire de Paris et de l'Ile-de-France*, 38 (1911), 224–248; Marc Bloch, "Cerny ou Serin," *Annales de la Société Historique et Archéologique du Gatinais*, 30 (1912), 157–160.

70 Marc Bloch, review of *The Making of Western Europe*, I, by C. R. L. Fletcher, *RSH*, 24 (1912), p. 417; Marc Bloch, review of *A History of Diplomacy in the International Development of Europe*, I-III, by David Hill, *RH*, 119 (1915), 389–393; Marc Bloch, review of *Histoire de la dîme ecclésiastique dans le royaume de France aux XIIe et XIIIe siècles*, by Paul Viard, *RSH*, 27 (1913), 374–375.

71 See also, AN AB^{XIX} 3830; AB^{XIX} 3831.

72 Marc Bloch, review of *Der Deutsche Staat*, p. 347. In his review of Febvre's *Histoire de Franche-Comté* Bloch suggested that much interesting work on the origins and decline of provincial patriotism remained to be done. Marc Bloch, review of *Histoire de Franche-Comté*, pp. 355–356.

73 Marc Bloch, "M. Flasch et les origines de l'ancienne France," *RSH*, 31 (1920), 152.

Cf. Gernet, "L'Approvisionnement d'Athènes," p. 383. For a discussion of the treatment of political phenomena by the Durkheimians, see Pierre Favre, "The absence of political sociology in the Durkheimian classifications of the social sciences," in Besnard, *The Sociological Domain,* pp. 199–216.

6 The University of Strasbourg as a center of disciplinary change

1 University of Strasbourg, *Université de Strasbourg: fêtes d'inauguration* (Strasbourg: Imprimerie Alsacienne, 1920), p. 32.

2 Gustave Lanson, "La Renaissance de l'Université française de Strasbourg," *RU,* 28 (1919), 325–326.

3 For a discussion of the German university see John E. Craig, *Scholarship and Nation Building: The Universities of Strasbourg and Alsatian Society, 1870–1939* (Chicago: University of Chicago Press, 1984), pp. 29–99; John E. Craig, "A mission for German learning: the University of Strasbourg," Ph.D., Department of History, Stanford, 1972.

4 Craig, *Scholarship and Nation Building,* pp. 59–60, 75.

5 For some of the proposals, see "Pour l'Université de Strasbourg," *La Vie Universitaire,* 2 (May, 1919), 15; A. Balz, "L'Université de Strasbourg," *La Vie Universitaire,* 3 (Aug., 1919), 5–7; Charles Andler, "La Rénovation présente des universités allemands," *La Vie Universitaire,* 3 (Feb., 1920) 9; Craig, *Scholarship and Nation Building,* p. 204.

6 Craig, *Scholarship and Nation Building,* pp. 205–207, 210.

7 University of Strasbourg, *Université,* pp. 31–32, 147–165.

8 This especially if one counts the numerous *chargés de cours* and *maîtres de conférence.* See *Index Generalis,* 1920 (Paris: Gauthier-Villars, 1921). The listing of twenty-two professors for Toulouse appears exaggerated; it seems likely that *maîtres de conférence* and *chargés de cours* were counted as professors. Cf. *Annuaire de l'Université de Toulouse pour l'année 1914–1915 et livret de l'étudiant* (Toulouse: Privat, 1914).

9 The other ones were Durkheim's chair at Bordeaux, subsequently held by Gaston Richard and Max Bonnafous, and two chairs at the Sorbonne initially held by Bouglé and Fauconnet. See Johan Heilbrun, "Les Métamorphoses du durkheimisme, 1920–1940," *RFS,* 26 (1985), 204–205.

10 See discussion and reproduction of Pfister's proposals in John Craig, "France's first chair of Sociology: a note on the origins," *Etudes Durkheimiennes,* no. 4 (Dec. 1979), 9–13.

11 See *Minerva* and *Index Generalis.* For a discussion of the controversy over the post of religious history, see Craig, *Scholarship and Nation Building,* pp. 218–219.

12 On Halbwachs, see Philippe Besnard, ed., *The Sociological Domain: The Durkheimians and the Founding of French Sociology* (Cambridge University Press and Paris: Editions de la MSH, 1983): Craig, "Sociology and related disciplines between the wars: Maurice Halbwachs and the imperialism of the Durkheimians," pp. 263–289; Besnard, "The epistemological polemic: François Simiand," p. 250.

13 Roussel contributed a few short reviews to the twelfth volume of the *Année Sociologique.* For Piganiol, see his article André Piganiol, "Qu'est-ce que l'his-

toire?" *Revue de Metaphysique et de Morale*, 60 (1955), 246. For Blondel on Durkheim, see Charles Blondel, *Introduction à la psychologie collective* (Paris: Colin, 1928); Charles Blondel, *Le Suicide* (Strasbourg: Librairie Universitaire d'Alsace, 1933); Charles Blondel, *La Conscience morbide* (Paris: Alcan, 1914). As an illustration of the strength of interest in Durkheimian sociology, the University of Strasbourg would match the University of Paris in having the highest numbers of collaborators to the new series of the *Année Sociologique* when it started in 1925, with the participation of Halbwachs, Roussel, Piganiol, Blondel, and Cahen. See title page, *AS*, n.s. 1, 1925.

14 Pierre George, "Max. Sorre," *Bulletin de la Section de Géographie* (CTHS), 81 (1975), 185. Best known for his theories on the evolution of landforms, W. M. Davis was a key figure in the professionalization of geography and the associated struggle for disciplinary independence both nationally and internationally. He was well liked by the Vidalians. See Robert P. Beckinsale, "W. M. Davis and American geography: 1880–1934," in Brian W. Blouet, ed., *The Origins of Academic Geography in the United States* (Hamden, CO: Archon Books, 1981), pp. 107–122; David N. Livingstone, *The Geographical Tradition: Episodes in the History of a Contested Enterprise* (Oxford: Blackwell, 1992), pp. 202–212. See also, DP letters of W. M. Davis to Albert Demangeon: Apr. 22, 1905; Mar. 26, 1911; Oct. 3, 1911; Dec. 5, 1911; July 14, 1912.

15 Christian Pfister, "La Faculté des Lettres de Strasbourg," *BFLS*, 1 (1922), 2.

16 The phrase is that of Andler: Andler, "La Rénovation présente," p. 9.

17 Henri Berr, "L'Esprit de synthèse dans l'enseignement supérieur," *RSH*, 32 (Jan.–July, 1921), 10.

18 Ibid., p. 11.

19 Cf. comments of Lot cited in Craig, *Scholarship and Nation Building*, p. 232.

20 Re frequent informal exchanges among faculty see Lucien Febvre, "Marc Bloch et Strasbourg: souvenirs d'une grande histoire," *Mémorial des années 1939–1945* (Paris: Belles Lettres, 1947), Publications de la Faculté des Lettres de l'Université de Strasbourg, fasc. 103, pp. 175, 181–182; Craig, *Scholarship and Nation Building*, pp. 230–233; Laurence K. Shook, *Etienne Gilson* (Toronto: Pontifical Institute of Mediaeval Studies, 1984), pp. 92–95.

21 The majority of those attending came from the Faculty of Letters. Representatives from the Faculté de Droit et des Sciences Politiques included Perrot, Champeux, Le Bras, Nast, Dusquesne and Niboyet, and from the Protestant Theology Faculty, Causse, Will, Seston, and Héring. Even the Faculties of Science and Medicine were occasionally represented as demonstrated by the participation of Gelma. Some of the professors such as Bloch and Cavaignac participated in all four at one time or another, and many were involved in at least two. See the *BFLS* for reports on the various meetings.

22 "Réunions de Samedi," *BFLS*, 1 (1922–1923), 106–107.

23 By Chipeaux, no doubt Febvre meant Ernest Champeaux.

24 HP: Conférences IV: Lucien Febvre to Henri Pirenne, n.d. (c. Oct., 1923).

25 This forum was begun with the help of Sylvain Lévi. Henri Baulig, "Lucien Febvre à Strasbourg," *BFLS*, 36 (1957–8), 180.

26 See, HP Activité Scientifique: Febvre to Henri Pirenne, Apr. 26, 1921.

27 See reports on the meetings for *histoire sociale* for Jan. 20, 1923; Nov. 24, 1923; and

Feb. 16, 1924: "Réunions de Samedi," *BFLS*: 1 (1922–1923), 152–153; 2 (1923–1924), pp. 88, 202. On meetings for *histoire littéraire,* see reports for Mar. 10, 1923; Apr. 21, 1923; Mar. 21, 1925; Dec. 8, 1928: "Réunions de Samedi," *BFLS*: 1 (1922–1923), 234–5, 277; 3 (1924–1925), 295; 7 (1928–1929), 199.

28 See the reports on the meetings for *histoire des réligions* for Dec. 2, 1922; Nov. 17, 1923; Mar. 15, 1924; Dec. 19, 1925; Apr. 24, 1926; Nov. 19, 1927; Feb. 25, 1928: "Réunions de Samedi," *BFLS*: 1 (1922–1923), 107; 2 (1923–1924), 47, 202–207; 4 (1925–1926), 85–86, 283–284; 6 (1927–1928), 40, 220–221. Halbwachs reviewed Davy's work at a meeting of the *histoire sociale* group on Mar. 17, 1923: *BFLS,* 1 (1922–1923), 193–194.

29 By 1935–1936 a new forum, the Centre d'Etudes de Philosophie de Droit et de Sociologie Juridique developed, perhaps in part because much of the life had gone out of *les réunions de samedi.*

30 See Marc Bloch and Lucien Febvre, "A Propos d'un concours," *AHES*, 6 (1934), 265–266; Marc Bloch, "Notes pour une révolution de l'enseignement," *Cahiers Politiques*, Aug. 3, 1943, pp. 17–24.

31 Berr, "L'Esprit de synthèse," pp. 12–13. Cf. Letters of Lucien Febvre to Henri Berr (before Easter 1920) and three years later (Autumn 1923) demonstrating that his initial euphoria had passed, both cited in Bertrand Müller, "Introduction," to Marc Bloch and Lucien Febvre, *Correspondance,* I (*La Naissance des Annales*), ed. Bernard Müller (Paris: Fayard, 1994), pp. xx, xxi.

32 Tabulated from Amicale de Secours des Anciens Elèves de l'ENS, *Annuaire*, 1940, pp. 156–164. See AN 61 AJ 207.

33 François Dosse, *L'Histoire en miettes: des "Annales" à la "nouvelle histoire"* (Paris: Editions La Découverte, 1987), pp. 13–14.

34 Baulig, "Lucien Febvre à Strasbourg," pp. 180–181. Later Febvre lost his enthusiasm for the project, and Bloch became the motivating force. As Febvre testified: "Vous vous rappelez que quand votre amicale insistance m'a convaincu qu'il fallait que je vous aide à mettre au pied les Annales dont j'avais à peu près abandonné le projet – je vous ai dit: 'Soit! mais naturellement l'affaire lancée , vous trouveriez bien que je reste un peu sous ma tente, et que je vous laisse mener la barque librement . . . Car d'autres études, pour l'instant m'attirent, et qui ne sont pas du ressort des Annales'." AN 318 MI 2, no. 176–177, letter from Lucien Febvre to Marc Bloch, Sept. 24, 1929. This letter suggests that Lyon's assertion that Febvre was the "paramount force" needs qualification. Cf. Bryce Lyon, "Introduction," to Lucien Febvre and Marc Bloch, *The Birth of Annales History: The Letters of Lucien Febvre and Marc Bloch to Henri Pirenne*, eds. Bryce and Mary Lyon (Brussels: Académie Royale de Belgique, Commission Royale d'Histoire, 1991), p. xix.

35 HP Activité Scientifique: letter of Lucien Febvre to Henri Pirenne, Apr. 26, 1921.

36 HP Activité Scientifique: letters of Lucien Febvre to Henri Pirenne, Apr. 26, 1921, and Marc Bloch to Henri Pirenne, Apr. 29, 1921. See also, HP Lucien Febvre, "Note sur l'organisation d'une Revue d'histoire et de sociologie économique," Dec. 5, 1921.

37 HP Activité Scientifique: letter of Marc Bloch to Henri Pirenne, Apr. 29, 1921; letter of Lucien Febvre to Henri Pirenne, Apr. 26, 1926; letter of Lucien Febvre to Henri Pirenne, Mar. 1, 1922.

38 HP Activité Scientifique: letter of Lucien Febvre to Henri Pirenne, May 31, 1922.

39 Lyon, "Introduction," to Febvre and Bloch, *The Birth*, p. xviii.

40 "Notes sur l'organisation," 5 Dec., 1921, in Febvre and Bloch, *The Birth*, p. 12; letters of Lucien Febvre to Henri Pirenne: late Mar. 1924; Apr. 24, 1924; in Febvre and Bloch, *The Birth*, letters 30 and 32, pp. 74–75, 78.

41 *Compte-Rendu du Ve Congrès International des Sciences Historiques: Bruxelles, 1923* (Nendeln, Liechtenstein, Kraus Reprint, 1972), pp. 291–294, 302–303. International Committee of Historical Sciences: *Bulletin*, 1 (1926–1929), pp. 189, 192–194, 337–344; *Bulletin*, 2 (1929–1930), pp. 355, 403–410.

42 Letter of Lucien Febvre to Henri Pirenne, May 27, 1928, Febvre and Bloch, *The Birth*; letter 43, p. 99. See also Müller, "Introduction," pp. xxiv-xxv, and Bloch and Febvre, *Correspondance*, 1, pp. 482–486 (letters of Marc Bloch to Eugène Schneider: Jan. 12, 1928; Jan 18, 1928; June 13, 1928; letters of Eugène Schneider to Marc Bloch: Jan. 13, 1928, Mar. 24, 1928.)

43 Re help of Demangeon see AN 318 MI 2, no. 1, letter of Lucien Febvre to Marc Bloch, n.d. (c. Feb. 1928). Lucien Febvre, *Combats pour l'histoire* (Paris: Colin, 1953), p. 382.

Using the Vidalians' major publisher was not without its problems, however. For example in 1929, Max Leclerc objected to a review of the *Géographie Universelle* by Febvre: "Je demandais simplement qu'il fût tenu compte du fait matériel que le Directeur et les collaborateurs de la 'Géographie Universelle' pourraient être peinés en constatant que c'est dans la maison même qui publie leurs travaux qu'auraient paru les critiques les plus sévères." AN 318 MI 1, no. 43, letter of Max Leclerc to Lucien Febvre, Mar. 14, 1929.

44 "Proceedings of the Sixth International Congress of Historical Sciences, Oslo, 1928," *Bulletin of the International Committee of Historical Sciences*, 2 (1929–1930), 105. AN 318 MI 1, no. 11–12, letter of Marc Bloch to Lucien Febvre, Aug. 22, 1928. Some German participation was now included. See, for example, letter of Lucien Febvre to Albert Thomas, Sept., 21, 1928, cited in Bertrand Müller, "'Problèmes contemporains' et 'Hommes d'Action' à l'origine des Annales," *Vingtième Siècle*, 35 (July–Sept., 1992), 85.

45 This journal began publication in 1928 but only lasted to 1932.

46 "Avis aux lecteurs," *RHM*, 1 (1926), 3–4. In their dealings with "la Maison Colin," Bloch and Febvre had pointed to the success of the foreign economic history journals and argued that within France a number of *bonnes volontés* were just waiting to collaborate with such a venture, this despite the competition for collaborators which they now faced from both French and foreign journals. Lucien Febvre, "minute de lettre," June, 1928 and Marc Bloch, letter of Jan. 12, 1928, both cited in Paul Leuilliot, "Aux origines des 'Annales d'histoire économique et sociale'," in *Mélanges en l'honneur de Fernand Braudel*, II (Toulouse: Privat, 1973), pp. 318–319.

47 Letter of Lucien Febvre to Max Leclerc, June, 1928, cited in Leuilliot, *Mélanges*, p. 319.

48 Letter of Marc Bloch to André Siegfried, Jan. 29, 1926, cited in Leuilliot, *Mélanges*, p. 318.

49 Lucien Febvre, "minute de lettre" cited in ibid. Cf. letter of Marc Bloch to André Siegfried, Jan. 29, 1929, cited in ibid.

50 The participation by Siegfried and Rist was to be largely symbolic. See Müller, "Introduction," pp. xxxvii-xxxix. Demangeon, on the other hand, served as both a

contact and a gatekeeper for the geographers with Bloch and Febvre deferring to his judgment on whether to approach various geographers.

51 AN 318 MI 2, no. 1–2, letter of Lucien Febvre to Marc Bloch, n.d. (c. Feb. 1929). See also letter of Max Leclerc, Mar. 6, 1928 cited in Leuilliot, *Mélanges*, p. 320.

52 Lucien Febvre, draft of a letter to Max Leclerc, Mar. 8, 1928, cited in Bloch and Febvre, *Correspondance*, I, pp. 9–10.

53 HP Febvre, "Note sur l'organisation."

54 Leuilliot, *Mélanges*, pp. 320–321; AN 318 MI 2, no. 4, letter of Lucien Febvre to Marc Bloch, n.d. (c. Feb. 1928); letter of Henri Pirenne to Marc Bloch, Apr. 8 (year not given but presumably 1928), cited in Leuilliot, p. 321.

    This journal has undergone several changes in title partly due to publication restrictions during the Second World War. In 1939, the title was changed to *Annales d'Histoire Sociale*; in 1942, the title became *Mélanges d'Histoire Sociale*; and in 1945, the journal was renamed *Annales d'Histoire Sociale*. In 1946, the title of *Annales: Economies, Sociétés, Civilisations* was first used, a title which was used until 1994, when it was changed to the present title, *Annales: Histoire, Sciences Sociales*.

55 Letter of Henri Pirenne to Marc Bloch, Apr. 8, 1928(?), cited in Leuilliot, *Mélanges*, p. 322. See also AN 318 MI 2, no. 33, letter of Lucien Febvre to Marc Bloch, n.d. (c. July, 1928) re Demangeon, Sion, and Levainville.

56 HP Febvre, "Note sur l'organisation"; letter of Lucien Febvre to Henri Pirenne, Mar. 1, 1922. Cf. title page *AS*, n.s. 1 (1925).

57 The international orientation of the Annales would also be evident in the second series of the *Annales Sociologiques* which replaced the *Année Sociologique* in 1934. Heilbrun, "Les Métamorphoses," p. 233.

58 HP Febvre, "Note sur l'organisation"; Les Directeurs (Marc Bloch and Lucien Febvre), "A Nos lecteurs," *AHES*, 1 (1929), 2. For a brief outline of the early organisation of the review see: Les Directeurs, "Au bout d'un an," *AHES*, 2 (1930), 1–3.

59 HP Febvre, "Note sur l'organisation"; letters from Lucien Febvre to Henri Pirenne, Dec. 4, 1921, Mar. 1, 1922 (Febvre and Bloch, *The Birth*, letters 3 & 6, pp. 7–21, 26–32). See also, AN 318 MI 1, no. 56, letter of Marc Bloch to Lucien Febvre, Aug. 21, 1929.

60 Henri Berr, *La Synthèse en histoire: son rapport avec la synthèse générale* (Paris: Albin Michel, 1953), p. 289.

61 Henri Berr, "Au bout de trente ans," *RS*, 1 (1931), 5. On the transformation of the journal and of its character see also, Müller, "Introduction," pp. xliii–xliv.

62 AN 318 MI 1, no. 2, letter of Marc Bloch to Lucien Febvre, May 11, 1928; AN 318 MI 2, no. 147–148, 157–161, 406, letters of Lucien Febvre to Marc Bloch, Aug. 26, 1929, Sept. 3, 1929, n.d. (c. Nov. 1933).

63 AN 318 MI 1, no. 78, letter of Marc Bloch to Lucien Febvre, Sept. 20, 1929.

64 The *Revue Historique* would also respond to some of these new interests but in a somewhat delayed fashion. See Alain Corbin, "La Revue Historique: analyse de contenu d'une publication rivale des Annales," in Charles-Olivier Carbonell and Georges Livet, eds., *Au berceau des Annales* (Toulouse: Presses de l'Institut d'Etudes Politiques de Toulouse), 1983, pp. 105–137. Lucette Le Van-Lemesle, "Une expérience de promotion de l'histoire economique: les Annales d'Histoire Economique et Sociale, 1929–1938," in Carbonell and Livet, eds., pp. 279–287.

65 Lanson, "La Renaissance"; André Hallays, "L'Université de Strasbourg: sa renaissance et son avenir," *Revue des Deux Mondes*, 53 (1919), 241–269.

66 For example, Pariset was replaced by Georges Lefebvre instead of by the Alsatian candidate, Braesch. Craig, *Scholarship and Nation Building*, p. 309. See also ibid., pp. 291–328.

67 Ibid., pp. 291–328.

68 See, for example, AN 318 MI 2, no. 323, letter of Lucien Febvre to Marc Bloch, Nov. 13, 1932; EB letter of Marc Bloch to Lucien Febvre, Oct. 16, 1933; AN 318 MI 1, no. 182, letter of Marc Bloch to Lucien Febvre, Mar. 12, 1936.

69 AN 318 MI 2, no. 450, letter of Lucien Febvre to Marc Bloch, n.d. (c. Mar. 1934). Letter of Lucien Febvre to Henri Pirenne, Apr., 1935, in Febvre and Bloch, *The Birth*, letter 80, p. 168.

70 Craig, *Scholarship and Nation Building*, pp. 325–326. By the end of the decade, however, the university found it increasingly difficult to fill the vacancies.

71 EB letter from Marc Bloch to Lucien Febvre, no. 11(35), c. early June, 1935.

72 See, for example, AN 318 MI 1, no. 94, 154, 182, letters of Marc Bloch to Lucien Febvre, Apr. 2, 1933; Apr. 13, 1935; May 12, 1936.

73 Heilbron, "Les Metamorphoses", p. 226; Olivier Dumoulin, "Changer l'histoire. Marché universitaire et innovation intellectuelle à l'époque de Marc Bloch," in Hartmut Atsma and André Burguière, *Marc Bloch aujourd'hui: histoire comparée et sciences sociales* (Paris: EHESS, 1990), pp. 87–104. See also, Dosse, pp. 26–27.

74 Bloch appeared to prefer a chair at the Collège de France over one at the Sorbonne because there he would be *plus libre.* Another incentive may have been that of the professors who were already there. For example, Febvre suggested that he was looking forward to "conférences Simiand-Febvre-Bloch" at the Collège de France. AN 318 MI 1, no. 189, letter of Marc Bloch to Lucien Febvre, Apr. 19, 1936; AN 318 MI 2, no. 447, letter of Lucien Febvre to Marc Bloch, n.d (c. Mar. 1934).

75 In an earlier competition when Bloch's candidacy was not so serious, he was up against Tonnelat and Vermeil, two other Strasbourg professors.

76 Dumoulin, "Changer l'histoire," pp. 87–104. Re anti-Semitism and Bloch's candidacy see for example, Fink, *Marc Bloch*, pp. 175–187.

7 Kings, serfs, and the sociological method

1 Carlo Ginsburg, "Prefazione," to Marc Bloch, *I Re Taumaturghi* (Turin: Giulio Einauldi, 1973), p. xv. At this point, Bloch still planned a principal thesis on the rural populations of the Ile-de-France, plans which changed by the autumn.

2 Marc Bloch, *Les Rois thaumaturges* (Paris: Gallimard, 1983), pp. xl–xli, 410.

3 AN 318 MI 1, no. 512, letter of Marc Bloch to Georges Davy, Apr. 9, 1924. See also AN 318 MI 1, no. 507, letter of Marc Bloch to Georges Davy, Oct. 4, 1923.

4 Marc Bloch, *Les Rois thaumaturges*, p. 21; AN 318 MI 1, letter of Marc Bloch to Lucien Febvre, Apr. 19, 1939.

5 Marc Bloch, *Les Rois thaumaturges*, pp. 51, 79, 86, 156. Bloch made a similar argument in a post-war article on war-time rumors in which he noted the importance of both the "fortuitous event" and the pre-existing "collective representations." Marc Bloch, "Réflexions d'un historien sur les fausses nouvelles de la

guerre," *RSH*, 33 (1921), 13–35, reproduced in Marc Bloch, *Mélanges historiques*, I (Paris: Fleury & EHESS, 1983), pp. 41–57.

6 François Simiand, "Méthode historique et science sociale," *RSH*, 6 (1903), 19.

7 Marc Bloch, *Les Rois thaumaturges*, p. 79.

8 Ibid., p. 21. In contrast to Jacques Le Goff's interpretation, Bloch did not claim to give "une explication totale" but only to examine that part relative to the milieu. See Jacques Le Goff, "Preface," in ibid., pp. xi-xii. For Bloch's investigation of the origins of the rite see chapters 1 & 2 in ibid.

9 Cf. Marc Bloch, "Réflexions d'un historien," pp. 48ff. where Bloch made similar comparisons between England and France.

10 Marc Bloch, *Les Rois thaumaturges*, p. 21.

11 Ibid., pp. 79ff., 171.

12 See, for example, ibid., pp. 57, 62, 156, 252, 294–5.

13 Ibid., p. 53.

14 Ibid.

15 Ibid., p. 52.

16 Ibid., pp. 55, 66ff.

17 His strong interest in religious phenomena during this period was also evident in three articles related to his research for this work. See Marc Bloch, "Saint Martin de Tours," *Revue d'Histoire et de Littérature Religieuses* (1921) as reproduced in Marc Bloch, *Mélanges historiques*, II (Paris: Fleury & EHESS, 1983), pp. 939–947; Marc Bloch, "La Vie de Saint-Edouard le Confesseur par Osbert de Clare," *Analecta Bollandia*, 41 (1923), 5–131 as reproduced in Marc Bloch, *Mélanges historiques*, II, pp. 948–984; Marc Bloch, "La Vie d'outre-tombe du roi Solomon," *Revue Belge de Philologie et d'Histoire*, 4 (1925), 349–377. See also his later review of a work on Saint Nicholas: Marc Bloch, "Ce qu'enseigne l'histoire d'un grand saint," *RS*, 7 (1934), 211–216; and his notes for a talk on Saint Martin which he gave at *la réunion de samedi, histoire des religions* in AN AB[xix] 3827.

18 As leaders became more skeptical of the effectiveness of the rite, the words *Dieu te guérit* changed to *Dieu te guérisse,* the subjunctive form indicating greater doubt. This was an argument not unlike the sort which Gernet had advocated. Marc Bloch, *Les Rois thaumaturges*, p. 399. Cf. Louis Gernet, "Authentes," *Revue des Etudes Grecques*, 22 (1909), 13–32 and Louis Gernet, *Recherches sur le développement de la pensée juridique et morale en Grèce* (Paris: Leroux, 1917), p. x.

19 Marc Bloch, *Les Rois thaumaturges*, p. 76.

20 See also letter of Marc Bloch to Henri Pirenne, June 30, 1933, in Marc Bloch and Lucien Febvre, *The Birth of Annales History: The Letters of Lucien Febvre and Marc Bloch to Henri Pirenne (1921–1935),* eds., Bryce and Mary Lyon (Brussels: Académie Royale de Belgique, Commission Royale d'Histoire, 1991), letter 72, p. 157. There Bloch writes of the importance of studying symbols with "ce sens du concret et ce goût de la simplicité que vous nous enseigne."

21 Marc Bloch, "Réflexions d'un historien sur les fausses nouvelles," as cited in *Mélanges*, p. 54.

22 For Bloch's use of these terms see Marc Bloch, *Les Rois thaumaturges*, pp. 17, 39, 51–53, 69, 86, 175, 250, 255–256, 390, 409.

23 Ibid., pp. 17, 41, 52, 110, 114, 154, 259, 345. See also the useful discussion in Le Goff, "Preface," to ibid., pp. xxvii-xxix.

24 Emile Durkheim, *De la division du travail social* (Paris: Alcan, 1926), pp. 46, 148. For a useful discussion of Durkheim's use of the terms "collective consciousness" and "collective representations" see Steven Lukes, *Emile Durkheim: His Life and Work, a Historical and Critical Study* (Harmondsworth: Penguin, 1973), pp. 4–8. According to Lukes, as Durkheim's interest turned to more advanced societies, he abandoned the concept of "collective consciousness."

25 Marc Bloch, *Les Rois thaumaturges*, pp. 69, 175.

26 See, for example, Marcel Mauss, "Rapports réels et pratiques de la psychologie et de la sociologie," *Journal de Psychologie Normale et Pathologique* (1924), reproduced in Marcel Mauss, *Sociologie et anthropologie* (Paris: PUF, 1950), pp. 288-291.

27 Emile Durkheim, *Les Règles de la méthode sociologique* (Paris: PUF, 1950), p. xvii; Paul Fauconnet and Marcel Mauss, "La Sociologie," in *La Grande Encyclopédie*, xxx (Paris: Société Anonyme de la Grande Encyclopédie, 1901), p. 171.

28 Emile Durkheim and Marcel Mauss, "De quelques formes primitives de classification: contribution à l'étude des représentations collectives," *AS*, 6 (1903), 1–72; Marcel Mauss with the collaboration of Henri Beuchat, "Essai sur les variations saisonnières des sociétés eskimos," *AS*, 9 (1906), 39–132; Emile Durkheim, "Représentations individuelles et représentations collectives," *Revue de Métaphysique et de Morale*, 6 (1898), reproduced in Emile Durkheim, *Sociologie et philosophie* (Paris: PUF, 1951), p. 43.

29 Marc Bloch, *Les Rois thaumaturges*, p. 409.

30 Ibid., pp. 106, 284, 312, 385.

31 Marc Bloch, review of *Introduction à la psychologie collective*, by Charles Blondel, *RH*, 160 (1929), 399, note 1.

32 Marc Bloch, *Les Rois thaumaturges*, p. xl. For references to his work as one of collective psychology, see AN 318 MI 1, no. 507, 512, letters of Marc Bloch to Georges Davy, Oct. 4, 1923 and Apr. 9, 1924.

33 Marc Bloch, review of *Introduction à la psychologie collective*, p. 399.

34 Ibid., p. 398.

35 See for example Charles Blondel, *Introduction à la psychologie collective* (Paris: Colin, 1928), pp. 199–201.

36 Cf. discussion in P.L. Orsi, "La storia delle mentalita in Bloch e Febvre," *Rivista de Storia Contemporanea*, 14 (1983), 374–375. Lévy-Bruhl, although he had not yet had time to examine Bloch's massive book, remarked to Bloch, "Le sujet de vos rois 'thaumaturges' m'intéresse au plus haut point. Je n'étudie la mentalité dite primitive que dans des sociétés aussi différentes que possible des nôtres: mais je suis reconnaissant à ceux qui étudient une mentalité analogue dans des régions et des temps accessibles à l'histoire, comme vous le faites. Il y aura là pour moi matière à des réflexions et des comparaisons précieuses." AN AB[xix] 3851, letter of Lévy-Bruhl to Marc Bloch, Apr. 8, 1924. On the relationships between Lévy-Bruhl and Bloch, see also Bloch, *Les Rois thaumaturges*, p. 421, note 1. Bloch's close friend, Georges Davy, was also significantly influenced by Lévy-Bruhl. See Georges Davy, *La Foi jurée* (Paris: Alcan, 1922).

37 See Marc Bloch, "Réflexions d'un historien," as cited in *Mélanges historiques*, p. 44. Despite his acknowledgment to Febvre, Bloch's use of psychology was still very different from Febvre's. See interesting discussion in François Dosse,

L'Histoire en miettes: des "Annales" à la "nouvelle histoire" (Paris: Editions La Découverte, 1987), pp. 77–88.

38 Maurice Halbwachs, review of *Les Rois thaumaturges*, by Marc Bloch, *AS*, n.s. (1923–1924), pp. 536–542. Halbwachs' rather critical review of *Les Rois thaumaturges* was in contrast to an earlier review in the *Année Sociologique* of a study of French coronation by the German historian Hans Scheuer. In 1912, the young Durkheimian E. Laskine had praised Scheuer for showing what the coronation rites meant for royalty as a whole and for demonstrating how far one could go with a study of formalism and juridical symbolism. The historian Paul Fournier had, however, criticized Scheuer for exaggerating the importance of magical and religious powers which the rite conferred. See reviews of Hans Scheuer, *Die rechtlichen Grundgedanken der französichen Königskrönung*: Edmond Laskine, *AS*, 12 (1909–1912), 460–465; Paul Fournier, *Journal des Savants*, n.s. 11 (1911), 116–120.

Another sociologist to whom one might turn to uncover reactions to *Les Rois thaumaturges* is Georges Davy. As Bloch wrote to Davy, he had asked Lévy-Bruhl for a review in his *Revue Philosophique*. Lévy-Bruhl in turn suggested that they use Davy begging off himself because of a busy schedule, a solution which Bloch told Davy suited him perfectly (AN 318 MI 1, no. 512, letter of Marc Bloch to Georges Davy, Apr. 9, 1924). However, such a review does not appear to have been published. Bloch also asked Berr to approach Granet for a review in the *Revue de Synthèse Historique*, but that review was to be written by Louis Rougier instead. Letter of Marc Bloch to Henri Berr, May 1, 1924, in Marc Bloch, *Ecrire La Société féodale: lettres à Henri Berr*, ed., Jacqueline Pluet-Despatin, (Paris: IMEC, 1992), letter 1, p. 29.

39 By contrast, E. F. Jacob, the English historian, wanted more on the healing powers of the Roman emperors. Lyn Thorndike, a history professor at Columbia, simply mentioned that "his method is almost always strictly historical and not that of the sociologist or student of folk-lore." See reviews of *Les Rois thaumaturges*: E.F. Jacob, *English Historical Review*, 40, 158, pp. 267–270; Lyn Thorndike, *American Historical Review*, 30, 3 (1925), p. 585.

40 Christian Pfister, review of *Les Rois thaumaturges*, *Journal des Savants*, 23 (1925), 110.

41 François Ganshof, review of *Les Rois thaumaturges*, *Revue Belge de Philologie et de l'Histoire*, 5 (1926), 611–615; Ch. Guignebert, review of *Les Rois thaumaturges*, *RH*, 148 (1925), 100–103.

42 R. Fawtier, review of *Les Rois thaumaturges*, *Moyen Age*, 36:27, 2nd ser. (1926), 241. AN AB[xix] 3851: letters from Henri Sée to Marc Bloch, n.d.; from Henri Pirenne to Marc Bloch, May 11, 1924; and from Gustave Lanson to Marc Bloch, May 11, 1924. Sée wrote, "Votre dernier chapitre, excellent, n'interessera pas seulement les historiens: folkloristes, psychologues, et sociologues y trouveront matière à réflexions." Pirenne praised Bloch's contribution as "de la plus haute valeur à la connaissance des idées politiques, religieuses et sociales."

In his review for the *Revue des Questions Historiques*, known for its Catholic and conservative orientation, de Croy attacked Bloch for his rationalism, irony and attempt to dismiss supernatural phenomena. J. de Croy, *Revue des Questions Historiques*, 7 (1925), 429–435.

43 Marc Bloch, "Le Popularité de toucher des écrouelles," *Moyen Age*, 37 (28th of 2nd ser.) (1927), 34.

44 AN AB^{xix} 3851, letters of Lucien Febvre to Marc Bloch, n.d. (c. Apr.–May 1924), and Maurice Prou to Marc Bloch, Apr. 9, 1924; Ernest Perrot, review of *Les Rois thaumaturges*, *Revue Historique du Droit Français et Etranger*, 4th ser., 6 (1927), 322–326; G. Dupont-Ferrier, review of *Les Rois thaumaturges*, *RU*, 34 (1925), 56. For other reviews see P. F. Fournier, review of *Les Rois thaumaturges*, *Bibliothèque de l'Ecole des Chartes*, 86 (1925), 192–194; Louis Rougier, "Les Rois thaumaturges d'après un ouvrage récent," *RSH*, 39 (1925), 95–106; S. Reinach, review of *Les Rois thaumaturges*, *Revue Archéologique*, 20 (1924), 280–1; Anon., review of *Les Rois thaumaturges*, *BFLS*, 2 (1923–1924), 259–260.

45 See, for example, Marc Bloch, "Les Annales Sociologiques," *AHES*, 7 (1935), 393 (re Mauss); Cf. Marc Bloch, "Le Développement de Paris depuis le milieu du XIXe siècle," *AHES*, 1 (1929), 435–436.

46 AN 318 MI 1, no. 500, letter of Marc Bloch to Georges Davy, Mar. 28, 1921.

47 Letter of Marc Bloch to Henri Berr, May 1, 1924, in Marc Bloch, *Ecrire*, letter 1, p. 27.

48 Marc Bloch, "Serf de la glèbe: Histoire d'une expression toute faite," *RH*, 136 (1921), 221–224.

49 Ibid., p. 242; Marc Bloch, "Servus glebae," *Revue des Etudes Anciennes*, 24 (1926), 352–358; Marc Bloch, "Les Colliberti: étude sur la formation de la classe servile," *RH*, 157 (1928), 1–48, 225–263; and Marc Bloch, "Collibertus ou Culibertus?" *Revue de Linguistique Romane*, 2 (1926), 16–24.

50 Marc Bloch, "Comment et pourquoi finit l'esclavage antique," *AESC*, 2 (1947), 37–44, 161–170. From Bloch's correspondence, it appears that this article was originally written in 1932 as a chapter for Bloch's unfinished work on the origins of the European economy for Berr, which makes his insistence on the importance of religious factors even more interesting. Letter of Marc Bloch to Henri Berr, Feb. 1, 1933, in Marc Bloch, *Ecrire*, letter 18, p. 66.

Re inability of serfs to enter religious orders see Marc Bloch, "Liberté et servitude personnelles au Moyen Age, particulièrement en France: contribution à une étude des classes," *Anuario de Histoira del Derecho Espanol*, (1933), 5–101, reproduced in Bloch, *Mélanges historiques*, I, p. 325. See also, Marc Bloch, "Classification et choix des faits en histoire économique," *AHES*, 1 (1929), 258.

51 Marc Bloch, "Liberté et servitude personnelles," as cited in *Mélanges*, I, p. 355. This article was originally given as a speech at the Semaine d'Histoire du Droit de Madrid in May, 1932. Bloch also discussed social class in his later works. See Marc Bloch, *Les Caractères originaux de l'histoire rurale française* (Paris: Colin, 1964), pp. 194–200; Marc Bloch, *La Société féodale* (Paris: Albin Michel, 1983), pp. 395–479; Marc Bloch, "Problèmes de classes en France et en Angleterre au Moyen Age," a talk given at Cambridge at Easter, 1939 found in AN AB^{xix} 3834.

52 Maurice Halbwachs, "Remarques sur la position du problème sociologique des classes," extract from *Revue de Metaphysique et de Morale* (1905), reproduced in Maurice Halbwachs, *Classes sociales et morphologie* (Paris: Edition du Minuit, 1972), p. 43. Bloch's files contain notes on this article in which he observed that it was somewhat confusing and marred by Halbwach's obsession with deciding whether or not a given question was sociological: AN AB^{xix} 3839.

53 Marc Bloch, "Une analyse de la vie économique," *RSH*, 52 (1931), 256. See also Bloch's notes on Simiand in AN AB^{xix} 3839.

54 Bloch, "Une analyse de la vie économique," p. 256. On the question of "materialistic" explanations, Bloch was also skeptical of explanations of medieval social structure which relied solely on economic factors, writing, "En un mot, l'institution seigneuriale n'est intelligible que comme un des éléments d'un système social fondé sur les relations de protection." Marc Bloch, "Classification et choix," p. 257.

55 Bloch, "Liberté et servitude personnelles," pp. 325, 348. On the confusion over who should be classified as a serf, see Marc Bloch, "Les Transformations du servage: à propos de deux documents relatifs à la région parisienne," in *Mélanges d'histoire offerts à Ferdinand Lot* (Paris: Champion, 1925), pp. 224–274; and Marc Bloch, "De la cour royale à la cour de Rome: le procès des serfs de Rosny-sous Bois," *Studi di Storia e Diritto in onore di E. Besta* (Milan: Dott. A. Giuffre, 1938), II, reproduced in Bloch, *Mélanges historiques*, I, pp. 452–461.

56 Marc Bloch, "Un problème d'histoire comparée: la ministérialité en France et en Allemagne," *Revue Historique de Droit Français et Etranger* (1928), 46–91, reproduced in *Mélanges historiques*, I, pp. 503–528. Cf. the emphasis Bloch put on *genre de vie* in his discussion of social class in Marc Bloch, *Les Caractères originaux*, pp. 194–197.

57 Marc Bloch, "Sur le passé de la noblesse française: quelques jalons de recherche," *AHES*, 8 (1936), 367. Cf. Henri Sée, *Les Classes rurales et le régime domanial en France au Moyen Age* (Paris: Giarad et Brière, 1901); Henri Sée, *Esquisse d'une histoire du régime agraire en Europe aux XVIIIe et XIXe siècles* (Paris: Giard, 1921). Bloch did not, however, totally reject a legal approach and had cautioned in a 1927 review against an over-reaction to *l'excès habituel*. Marc Bloch, "Quelques études récentes sur l'histoire économique et sociale de l'ancienne France," *RSH*, 43 (1927), 98.

58 Marc Bloch, "Sur le passé de la noblesse," p. 370.

59 Marc Bloch, "Mémoire collective, tradition et coutume à propos d'un livre récent," *RSH*, 40 (1925), 82. See also Marc Bloch, "Sur l'étude des classes, au Moyen Age," *AHES*, 7 (1935), 214–215.

60 Re "vicissitudes" of social classification, see for example, letter of Marc Bloch to Henri Pirenne, Feb. 16, 1933, in Febvre and Bloch, *The Birth of Annales History,* letter 69, p. 150.

61 Marc Bloch, "A propos d'une étude sur les liens seigneuriaux en Flandre: problèmes de méthode," *AHES*, 9 (1937), 304.

62 Other examples of social classifications would include *les confrèries,* the villages, and the concept of "collective organism" as used by American sociologists. Marc Bloch, "Les Confrèries," *AHES*, 1 (1929), 590–591; Marc Bloch, "Une nouvelle théorie sur l'origine des communes rurales," *AHES*, 1 (1929), 587–589; Marc Bloch, "Les Sociétés rurales," *AHES*, 4 (1932), 475. Generally suspicious of explanations relying on ethnicity, Bloch did not entirely rule out a study of ethnic groups. See, for example, Marc Bloch, "Les Origines ethniques d'une classe," *AHES*, 2 (1930), 596; and Marc Bloch, "Un essai de psychologie collective," *AHES*, 10 (1938), 175–176.

63 See, for example, Nancy Green, "L'Histoire comparative et le champ des études migratoires," *AESC*, 45 (1990), 1336.

64 Marc Bloch, "Pour une histoire comparée des sociétés européennes," *RSH*, 46 (1928), p. 49.

65 Letters of Marc Bloch to Henri Berr, Oct. 1, 1928; Nov. 28, 1928 in Marc Bloch, *Ecrire*, letters 9, 12, pp. 52, 54.

66 Marc Bloch, "Pour une histoire comparée," p. 17. Bloch himself made this link, see p. 20. See also the abstract of Bloch's talk in VIᵉ Congrès international des Sciences Historiques, *Résumés des communications présentés au Congrès, Oslo, 1928* (Le Comité Organisateur du Congrès, 1928), pp. 119–121.

67 Marc Bloch, "Pour une histoire comparée," p. 19.

68 Ibid.

69 Cf. Marc Bloch, review of *The Medieval City*, by G. C. Coulton, *Revue Critique d'Histoire et de Littérature*, n.s. 93 (1926), 281–283. See also, Marc Bloch, review of *La Réforme agraire en Europe*, by A. Wauters, *AHES*, 1 (1929), 102–3.

70 See Meitzen's classic work, August Meitzen, *Siedlung und Agrarwesen der Westgermanen und Ostgermanen, der Kelten, Römer, Finnen, und Slaven*, I–III (Berlin: W. Hertz, 1895).

71 Marc Bloch, "Pour une histoire comparée," p. 45.

72 Antoine Meillet, *La Méthode comparative en linguistique historique* (Oslo: Aschehouge & Co., 1925), pp. 1–11. Bloch, "Pour une histoire comparée," p. 16, note 1. On Meillet and the Durkheimians, see Joseph Vendryes, "Antoine Meillet," in Thomas A. Sebeok, ed., *Portraits of Linguists* (Bloomington, IN: Indiana University Press, 1963), II, p. 223.

73 Marc Bloch, "Pour une histoire comparée," p. 41.

74 This goes contrary to the assertion by the Hills that Bloch derived his model from Meillet but demonstrated a lack of rigor in following it. Cf. Alette Olin Hill and Boyd Hill, Jr., "Marc Bloch and comparative history," *The American Historical Review*, 85 (1980), 829, 833.

Despite the essential differences between Bloch and Meillet in both aims and methods, Bloch would use an approach much like that of Meillet when interpreting linguistic evidence. For example, he tried to show that by comparing *culvert* to equivalents in Spanish and Italian one could note common origins pointing to pre-Roman roots. See Marc Bloch, "Les Colliberti."

75 Marc Bloch, "Pour une histoire comparée," p. 35, note 1.

76 Emile Durkheim, review of *La Tradition romaine sur la succession des formes du testament devant l'histoire comparative*, by Edouard Lambert, *AS*, 5 (1902), 376.

77 Ibid., p. 375.

78 Emile Durkheim, *Les Formes élémentaires de la vie religieuse*, 4th ed., (Paris: PUF, 1960), p. 133. Georges Davy made similar criticisms of Frazer in his 1921 review: Georges Davy, "A propos de l'Ancien Testament: une nouvelle contribution de M. Frazer à l'histoire comparative des institutions," *RSH*, 32 (1921), p. 96.

79 See Durkheim, *Les Formes*, pp. 135–136.

80 Durkheim, *Les Règles de la méthode sociologique*, p. 137. Cf. Emile Durkheim, "Représentations individuelles et représentations collectives," in Durkheim, *Sociologie et philosophie*, p. 1.

81 Marc Bloch, "Pour une histoire comparée," p. 15.

82 For an interesting discussion of this point see J. Ambrose Raftis, "Marc Bloch's

comparative method and the rural history of mediaeval England," *Mediaeval Studies*, 24 (1962), 350.

83 Henri Pirenne, "De la méthode comparative en histoire,"

83 Henri Pirenne, "De la méthode comparative en histoire," *Compte-Rendu du Ve Congrès International des Sciences Historiques* (Brussels, 1923) (Nendeln, Liechtenstein: Kraus Reprint, 1972), pp. 19–32; Henri Berr, "Le Ve Congrès International des Sciences Historiques et la synthèse en histoire," *RSH*, 35 (N.S. 9) (1923), 10–11; Henri Sée, "Remarques sur l'application de la méthode comparative à l'histoire économique et sociale," *RSH*, 36, n.s. 10 (1923), 37–46; Charles V. Langlois, "The comparative history of England and France during the middle ages," *English Historical Review*, 5 (1890), 261. As Bloch confided to Berr, he cited Sée's article "plutôt par courtoisie et pour l'auteur et aussi pour la périodique, que pour le travail lui-même." Letter of Marc Bloch to Henri Berr, Oct. 1, 1928 in Bloch, *Ecrire*, letter 11, p. 52.

Bloch also directed the reader's attention to a lengthy article by Davaillé but noted that it was "conçus dans un esprit différent de l'étude qu'on va lire." Marc Bloch, "Pour une histoire comparée," p. 15. Cf. Louis Davaillé, "La Comparaison et la méthode comparative en générale," *RSH*, 27 (1913), 9–33, 217–257.

84 Marc Bloch, "Comparaison," *Bulletin du Centre International de Synthèse* (supple. to *RSH*) 9 (1930), 35. Initially Berr had asked Sée to contribute this article but as his commission was not satisfied with the results, he approached Bloch to replace him. Marc Bloch, *Ecrire*, p. 53 (note 3 to letter from Marc Bloch to Henri Berr, Oct. 1, 1928, letter 11).

85 Ibid., p. 39.

86 See "Séance du Mercredi 8 Janvier, 1930," *Bulletin du Centre International de Synthèse* (supple. to *RSH*), 9 (June 1930), 15–19. Berr by this time was preparing for his new journal *Revue de Synthèse* which was to embrace all of science. Bloch's "projet d'article" was soon followed by another more general one: R. Bouvier, "Comparison," *RS*, 1 (1931), 51, note 1.

87 However, in the discussion following Bloch's paper in January, 1930, Bloch had indicated that to speak of "Japanese feudalism" made him uncomfortable: "Séance du mercredi 8 janvier," p. 18.

88 Marc Bloch, "Un essai d'histoire comparée: Europe occidentale et Japon," *AHES*, 2 (1930), 136–137; Marc Bloch, "Féodalité et vassalité japonaise," *AHES*, 5 (1933), 598; Marc Bloch, "Un voyage à travers l'histoire comparée," *RS*, 3 (1933), 324–325; Marc Bloch, "Histoire, doctrine économique, sociologie," *AHES*, 6 (1934), 510. Marc Bloch, "Les Féodalités: une enquête comparative," *AHES*, 8 (1936), 591; Marc Bloch, "Problèmes d'histoire comparée," *AHS*, 1 (1939), 438–440.

Bloch's comparisons have recently been criticized precisely for this Eurocentrism, which he himself admitted, but did not see how to escape. See Hartmut Atsma and André Burguière, eds., *Marc Bloch aujourd'hui: histoire comparée et sciences sociales* (Paris: EHESS, 1990): Evelyn Patlagean, "Europe seigneurie, féodalité. Marc Bloch et les limites orientales d'un espace de comparaison," pp. 279–298; Michel Cartier, "Les Historiens chinois, du marxisme au comparatisme. L'exemple du féodalisme," pp. 299–305; Lucette Valensi, "Retour d'Orient. De quelques usages du comparatisme," pp. 307–316.

89 EG letter of Marc Bloch to Etienne Gilson, Dec. 28, 1933; EB letters of Marc

Bloch to Lucien Febvre, Dec. 19, 1933; Dec. 24, 1933; Mar. 22, 1934. Febvre in turn suggested that Bloch emphasize economic history. AN 318 MI 2, no. 343, letter of Lucien Febvre to Marc Bloch, n.d. (c. late 1933). See also discussion in Olivier Dumoulin, "Changer l'histoire. Marché universitaire et innovation intellectuelle à l'époque de Marc Bloch," in Atsma and Burguière, eds., *Marc Bloch,* pp. 88–89.

90 Marc Bloch, "Projet d'un enseignement d'histoire comparée des sociétés européennes," Imprimerie des Derniers Nouvelles de Strasbourg, Dépot Légal, Département du Bas Rhin, 1934. (BN 8° G. Piece 2023), 16 pp. Cf. EB Dossier: Histoire des doctrines économiques, back of notes re "le mathématique et la science sociale." There Bloch argued that when undertaking a comparative study of feudal society, one should first form a clear idea of its manifestation in a particular "régime social" as a reference point.

91 See also his comments to Henri Pirenne on the difficulties of writing comparative history, "Le terrible est que toutes les fois qu'on cherche à dresser un questionnaire ... les trois quarts du temps au moins il est impossible de trouver la réponse." Marc Bloch to Henri Pirenne, July 21, 1935, in Febvre and Bloch, *The Birth,* letter 84, p. 175.

92 Johan Heilbron, "Les Métamorphoses du durkheimisme, 1920–1940," *RFS,* 26 (1985), 203–236.

93 Ibid., pp. 211–236.

94 AN 318 MI 2, no. 346, letter of Lucien Febvre to Marc Bloch, c. June 1933.

95 For Bloch's interest in memory see also letter of Marc Bloch to Henri Pirenne, Aug. 30, 1934, in Marc Bloch, *Ecrire,* letter 75, p. 161.

96 Marc Bloch, "Mémoire collective, tradition et coutume, à propos d'un livre récent," pp. 73–83; Marc Bloch, "Le Développement de Paris depuis le milieu du XIXᵉ siècle," p. 436. Bloch's 1925 review which appeared in the *Revue de synthèse historique* was done at Bloch's request. See letter of Marc Bloch to Henri Berr, June 18, 1925, in Marc Bloch, *Ecrire,* letter 7, p. 46.

97 Marc Bloch, "Un symptôme social: le suicide," *AHES,* 3 (1931), 592.

98 Marc Bloch, "La Répartition des dépenses comme caractère de classes," *AHES,* 7 (1935), 83–86.

99 Marc Bloch, "Les Formes matérielles de la vie sociale," *AHS,* 1 (1939), 316.

100 Marc Bloch, "Le Développement de Paris," p. 436.

101 See also Bloch's letter to Pirenne, in which he criticizes Durkheimian sociology for trying to separate sociology and history. There he wrote, "Nous devons beaucoup – et j'ai moi-même largement conscience de ma dette – aux efforts des sociologues, de l'école de Durkheim particulièrement. Mais je crois que sur un point ils ont vu faux. L'existence, côte à côte, d'une histoire de d'une sociologie me paraît la plus artificielle des constructions. C'est un reste de l'état métaphysique." Letter of Marc Bloch to Henri Pirenne, n.d. in Febvre and Bloch, *The Birth,* letter 59, p. 131. Contrary to Bryce Lyon's interpretation, what Bloch appears to be saying here is that history and sociology were too close to separate not that their methods were fundamentally different. See, for example, Bloch's remarks to Berr in his letter of Oct. 1, 1928, in Bloch, *Ecrire,* letter 11, p. 52. Cf. Bryce Lyon, "Introduction" to Febvre and Bloch, *The Birth,* pp. xli-xlii.

102 Marc Bloch, "Un symptôme sociale," pp. 591–592.

103 AN 318 MI 1, no. 111, letter of Marc Bloch to Lucien Febvre, May 9, 1933. Febvre, in fact, did write most of the reviews for the *Annales*, a solution which Bloch endorsed perhaps because Febvre as the senior partner was responsible for most of their contacts with Simiand.

104 AN 318 MI 1, no. 158, letter of Marc Bloch to Lucien Febvre, Apr. 18, 1935.

105 Marc Bloch, "Le Salaire et les fluctuations économiques à longue période," *RH*, 173 (1934), 4. The sociologists, by contrast, would continue to insist on an essential difference between the two fields. See discussion in Gérard Noiriel, "Pour une approche subjectiviste du social," *AESC*, 44 (1989), 1439.

106 Marc Bloch, "Une analyse de la vie économique," *RSH*, 52 (1931), 253–256.

107 Marc Bloch, "Le Salaire," pp. 28, 30.

108 Ibid., p. 30. Related to this discussion was Bloch's objection to Simiand's refusal to examine money in terms of its "valeur d'anticipation." According to Bloch, money, viewed in this way, expressed people's desires. See ibid., pp. 24–25; Marc Bloch, "La Monnaie réalité sociale," *AHES*, 8 (1936), 307.

109 Marc Bloch, "A travers l'histoire des prix et des monnaies," *RS*, 11 (1936), 237.

110 Marc Bloch, "Le Salaire," p. 8. Cf. Bloch's reviews of works by Elie Halévy and Henri Lévy-Bruhl: Marc Bloch, "Un peuple, une crise," *AHES*, 5 (1933), 430–431; M. F. (Marc Bloch), "Sur l'histoire des entreprises et des placements," *AHS*, 1 (1942), 94–95.

111 Marc Bloch, review of *Année Sociologique*, n.s. 1, *RH*, 155 (1927), 176; Marc Bloch, review of *Annales Sociologiques*, *AHES*, 7 (1935), 393. He observed that the second series of the *Année* was characterized by "un certain assouplissement dans les formules" and that the *Annales Sociologiques* begun in 1934 took on "une forme dorénavant 'plus libre et plus souple'"; Marc Bloch, review of *Annales Sociologiques*; Marc Bloch, "La Sociologie et le passé du droit," *AHES*, 8 (1936), 458; Marc Bloch, "Une equipe de sociologues," *AHES*, 7 (1935), 483, and Marc Bloch, review of *Annales Sociologiques*. On the *chercheurs* and *enseignants universitaires* see Heilbron, "Les Métamorphoses," pp. 224–225. According to Heilbron, Series A and to a lesser degree C were the outlets for the *enseignants universitaires* and only Series A attracted many young collaborators.

8 Reflections on the geographical approach and on the agrarian regime

1 On inter-war geography in France, see Numa Broc, "Les Séductions de la nouvelle géographie," in Charles-Olivier Carbonell and Georges Livet (eds.), *Au berceau des Annales* (Toulouse: Presses de l'Institut d'Etudes Politiques de Toulouse, 1983), pp. 247–263; Numa Broc, "Homo géographicus: radioscopie des géographes français de l'entre-deux guerres," *AG*, 102 (1993), 225–254; Paul Claval, "Le Rôle de Demangeon, de Brunhes, et de Gallois dans la formation de l'Ecole française des années 1905–1910," in Paul Claval, *Autour de Vidal de la Blache: la formation de l'école française de géographie* (Paris: CNRS, 1993), pp. 149–158.

2 Lucien Febvre, *La Terre et l'évolution humaine* (avec le concours de Lionel Bataillon) (Paris: La Renaissance du Livre, 1922), I, section 1 of Henri Berr's series "l'Evolution de l'Humanité." See, for example, the debate over Febvre's book which took place at Strasbourg in *une réunion de samedi* between Halbwachs, Febvre, Baulig, and Piganiol. "Réunions de samedi: Samedi, 20 janvier (histoire sociale),"

BFLS, 1 (1922–1923), 153. According to Febvre, he took on this volume only after Berr had received "Le refus successif de tous les géographes." DP Lucien Febvre to Albert Demangeon, Mar. 5, 1912.

3 Marc Bloch, review of *La Terre et l'évolution humaine*, by Lucien Febvre, *RH*, 145 (1924), 236.

4 Bloch, review of *La Terre*, p. 239.

5 Ibid. On Febvre's high esteem for Vidalian geography, which contrasted with Bloch's more critical assessment, see also letter of Lucien Febvre to Henri Pirenne, Feb. 16, 1924, in Lucien Febvre and Marc Bloch, *The Birth of Annales History: The Letters of Lucien Febvre and Marc Bloch to Henri Pirenne (1921–1935)*, eds. Bryce and Mary Lyon (Brussels: Académie Royale de Belgique, Commission Royale d'Histoire, 1991), letter 28, p. 69. There, Febvre wrote of "'l'esprit géographique' (qui n'est en réalité qu'une des formes de l'esprit historique)."

6 Maurice Halbwachs, review of "Le Problème de la géographie humaine à propos d'ouvrages récents," *AS*, n.s., 1 (1923–1924), 908.

7 On Febvre and *la géographie modeste*, see Yves Lacoste, "A bas Vidal, Viva Vidal," *Hérodote*, 16 (1979), 78–80.

8 AN AB^xix 3851, letter of Henri Sée to Marc Bloch, n.d.

9 Henri Hauser, review of *La Terre*, *Revue Critique de l'Histoire et de Littérature*, n.s. 91 (1924), 92–99; Albert Demangeon, review of *La Terre*, *AG*, 32 (1923), 165–170; Camille Vallaux, review of *La Terre* (and of two other works), *Mercure de France*, 161 (Jan. 7, 1923), 204–207; Camille Vallaux, "La Géographie humaine est-elle légitime et possible," *La Géographie*, 39 (1923), 461–463. For Hauser's positions on geography, see also H. Hauser, *L'Enseignement des sciences sociales* (Paris: Librairie Marescq aîné, 1903); H. Hauser and Joseph Fèvre, *Régions et pays de France* (Pairs: Alcan, 1909).

10 For the early planning of the series see DP: letter of Lucien Gallois to Albert Demangeon, July 8, 1908; letters of Vidal de la Blache to Albert Demangeon, May 22, 1909 and June 12, 1911; "Géographie Universelle publiée sous la direction de P. Vidal de la Blache et L. Gallois" (Paris: Colin, Mar. 1927), (promotional off-print for the *Géographie Universelle*) found in DP Dossier – Compte rendus de ses ouvrages, etc.

For a comparison of this *géographie universelle* and those which preceded it, see Robert Ferras, *Les Géographies Universelles et le monde de leur temps* (Montpellier: GIP reclus, 1989), Collection Reclus Modes d'Emploi no. 14.

11 Marc Bloch, review of *Iles Britanniques* (*Géographie Universelle*, I), by Albert Demangeon, *BFLS*, 6 (1927–8), 162. By contrast, Febvre wrote, "Or les pages que M. Demangeon a consacrées à retracer, d'ensemble, l'histoire du peuplement et du développement historique des Iles Britanniques sont d'un intérèt puissant et trahissent sa maîtrise plus complètement encore que toutes les autres." Lucien Febvre, *Pour une histoire à part entière* (Paris: EHESS, 1982), 84, 86. It appears, however, that Febvre would have written a more critical review of the *Géographie Universelle* for the *Annales*, but that Max Leclerc objected, given that Colin published the series, AN 318 MI 1, no. 43, letter of Max Leclerc to Lucien Febvre, Mar. 14, 1929.

12 Marc Bloch, review of *Belgique, Pays Bas, Luxembourg* (*Géographie Universelle*, II), by Albert Demangeon, *BFLS*, 6 (1927–8), 250–1; Marc Bloch, review of

Géographie Universelle, XV (Denis), 14 (Sorre), & 9 (Sion), *BFLS*, 7 (1928–9), 287–289; Marc Bloch, review of *Méditerranée: Peninsules méditerranéennes*, part 1, by M. Sorre and J. Sion, *Géographie Universelle*, VII, *BFLS*, 14 (1935–6), 112–113; EB: letter of Marc Bloch to Lucien Febvre, Nov. 7, 1933 (for Bloch's comments on Zimmerman's volume).

13 For example, in his review of Febvre's *La Terre et l'évolution humaine* Bloch remarked: "Dans sa conclusion, M. Febvre, exposant à grands traits le programme de la géographie historique, constate que cette science, au vrai sens du mot, n'existe pas encore. Cette dernière remarque n'est que trop vraie. Puissent les conseils de M. Febvre trouver quelque écho parmi les travailleurs!" Marc Bloch, review of *La Terre*, p. 240. Bloch made a similar observation as late as 1933 writing, "la géographie historique est une science qui n'a même pas atteint l'enfance, une science à faire." Marc Bloch, "L'Auvergne," *AHES*, 5 (1933), 318.

14 Marc Bloch, "Un atlas historique provincial," *AHES*, 5 (1933), 392. See also his derogatory reference to the "circuits où l'ont traînée Longnon et sa séquelle" in a letter to Lucien Febvre. Letter of Marc Bloch to Lucien Febvre, Nov. 10, 1926, in Marc Bloch and Lucien Febvre, *Correspondance*, I (*La Naissance des Annales: 1928–1933*), ed. Bertrand Müller (Paris: Fayard, 1994), p. 469.

15 AN 318 MI 1, no. 29, letter of Marc Bloch to Lucien Febvre, Sept. 27, 1928.

16 He pushed Febvre, for example, to review the plans for an historical atlas of France, writing to him, "Vous seul pouvez faire bonne besogne et empechiez qu'il ne s'en fasse de la mauvaise." AN 318 MI 1, no. 36, letter of Marc Bloch to Lucien Febvre, Oct. 2, 1928. See also, EB letter of Marc Bloch to Lucien Febvre, Oct. 4, 1933, re exhibition of historical maps at Warsaw.

17 Marc Bloch, "Une haute terre: l'Osians d'autrefois et d'aujourd'hui," *RSH*, 50 (1930), 71, note 1.

18 Marc Bloch, "Pour mieux comprendre le passé d'une partie du monde," *AHES*, 8 (1936), 584.

19 Ibid.; Marc Bloch, "En Angleterre: l'histoire et le terrain," *AHES*, 9 (1937), 208–210.

20 Marc Bloch, "Géographie et politique," *RS*, 11 (1936), 267–268. Bloch's conception of historical geography was also revealed in his reviews of historical atlases. There he argued that rather than attempting to convey *une totalité,* a concept *vidé de sens,* such atlases should include thematic maps. He longed for a true historical atlas of France: "un atlas, s'entend, qui ressorte à la fois de la géographie véritable et de l'histoire, au sens profond du mot, et ne se contente point de porter sur des cartes sans relief des frontiers parfois imaginaires." Marc Bloch, "Une erreur," *AHES*, 9 (1937), 384–5. See also Marc Bloch, "Géographie historique," *AHES*, 3 (1931), 597; Marc Bloch, "Bulletin historique: Histoire d'Allemagne," *RH*, 163 (1930), 366–7; Marc Bloch, "Un atlas historique provincial," 388–392; Marc Bloch, "Encore les cartes de frontière," *AHES*, 6 (1934), 290–1; Marc Bloch, "Une carte historique," *AHES*, 6 (1934), 582–584; Marc Bloch, "Réflexions sur un atlas historique scolaire," *AHES*, 6 (1934), 495–497; letter of Marc Bloch to Lucien Febvre, Nov. 10, 1926, in Bloch and Febvre, *Correspondance*, I, pp. 469–470.

21 Bloch made similar observations in his reviews of works on dwelling types and on urban and commercial geography. For example, he wrote to Febvre that Demangeon's address on "habitation" given at a folklore conference was poor, "il

s'est borné à répéter son viel article des *Annales* sur la classification des formes de maison, sans essai d'interpretation, sans un mot d'humain ou de social. Un cadre, et c'est tout." AN 318 MI 1, no. 199, letter of Marc Bloch to Lucien Febvre, Sept. 20, 1937. Bloch was, however, impressed by the following article which adopted an historical interpretation of "habitation": J. Célerier and E. Laoust, "L'Habitation dans le Moyen Atlas est mal adapté aux conditions naturelles," *Compte-rendus du Congrès International de Géographie, Varsovie, 1934*, III (Warsaw: Kasa Im. Mianowskiego, 1937), pp. 416–419. (See Bloch's copy marked *très important* in AN AB^{xix} 3832.)

See also Marc Bloch, "Champs et villages," *AHES*, 6 (1934), 482; Marc Bloch, "Une géographie commerciale," *RSH*, 48 (1929), 125–126; Marc Bloch, "Une ville et son milieu," *AHES*, 4 (1932), 600–601; Marc Bloch, "Un village français et l'après-guerre," *AHES*, 5 (1933), 321; Marc Bloch, "Une ville, une bourgeoisie," *AHES*, 8 (1936), 91–92; Marc Bloch, "Deux villes du Massif Central," *AHES*, 8 (1936), 505–506.

22 Marc Bloch, "Une enquête sur l'habitat rurale," *AHES*, 1 (1929), 421.

23 Marc Bloch, "Les Paysages agraires: Essai de mise au point," *AHES*, 8 (1936), 265. For similar observations, see also Marc Bloch, "Régions naturelles et groupes sociaux," *AHES*, 4 (1932), 489–493; Marc Bloch, "Champs et villages," 482–483; Marc Bloch, "Défrichement et habitat," *AHES*, 5 (1933), 319–320; Marc Bloch, "Paysages agraires du Nord," *AHS*, 3 (1941), 159–160.

24 See, for example, Marc Bloch, "Champs et villages," p. 481. For Meitzen's work, see August Meitzen, *Siedlung und Agrarwesen der Westgermanen und Ostgermanen, der Kelten, Römer, Finnen, und Slaven*, I-III (Berlin: W. Hertz, 1895). Due to fears of racial explanations, Bloch was very wary of ethnic studies; he did not, however, entirely rule out a possible influence of ethnic factors. See, for example, Marc Bloch, "Les Origines ethniques d'une classe," *AHES*, 2 (1930), 596; Marc Bloch, "Un essai de psychologie collective," *AHES*, 10 (1938), 175–176; Marc Bloch, "La Société du Haute Moyen Age et ses origines," *Journal des Savants*, (1926), 408–409.

25 Marc Bloch, "La Vie rurale: problèmes de jadis et de naguère," *AHES*, 2 (1930), 106–109. For a later review of Gradmann, see Marc Bloch, "Dans les steppes du Proche Orient," *AHES*, 10 (1938), 77–78. See also Bloch's notes on Gradmann's *Suddeutschland* challenging his interpretation of the Waldhufendorf, in AN AB^{xix} 3833. Bloch had much less praise for some of the French work. See Marc Bloch, "La Vie rurale," p. 111 re the works of Yvonne Bezard and Alphonse Schmidt; and Marc Bloch, "Les Champs et villages," pp. 182–183 re P. Déjean. That Bloch found this question, as well as that of *habitation,* to be of considerable interest is demonstrated by his notes on geographical works in AN AB^{XIX} 3832.

For an interesting commentary on Gradmann by a French geographer, see L. Champier, "A propos de l'oeuvre de Robert Gradmann (1865–1950): méthodes de recherche en géographie agraire," in *Annales Universitatis Saraviensis. Philosophie*, I (1952), pp. 190–202.

26 Marc Bloch, "Une monographie géographique: les pays du Rhône moyen," *AHES*, 1 (1929), 607; Marc Bloch, "Economie française: monographies géographiques," *AHES*, 1 (1929), 136.

27 Marc Bloch, "Economie française," p. 137.

28 Marc Bloch, "Une monographie géographique," p. 607.

29 Marc Bloch, "Champs et villages," p. 471.

30 Marc Bloch, "Quelques études récentes sur l'histoire économique et sociale de l'Ancienne France," *RSH*, 43 (1927), 95–99. On Sclafert, see also letter of Marc Bloch to Lucien Febvre, Nov. 10, 1926, Bloch and Febvre, *Correspondance*, I, p. 470.

31 For Bloch's comments on the treatment of social and geographical phenomena in the regional monographs see Marc Bloch, "Quelques études récentes sur l'histoire économique et sociale de l'Ancienne France" (Thérèse Sclafert); Marc Bloch, "Une monographie géographique," pp. 606–611 (Daniel Faucher); Marc Bloch, "Economie française," pp. 134–137 (Georges Chabot and Madeleine Bassère); Marc Bloch, "Une haute terre: l'Osians," pp. 71–78 (André Allix); Marc Bloch, "Régions naturelles et groupes sociaux," pp. 493–501 (André Meynier and Henri Cavaillès); Marc Bloch, "Un carrefour: la porte de Bourgogne," *RSH*, 53 (1933), 83–84 (André Gibert); Marc Bloch, "L'Auvergne," *AHES*, 15 (1933), 317–318 (Philippe Arbos); Marc Bloch, "Champs et villages," pp. 470–472 (Théodore Lefebvre); Marc Bloch, "En Corse," *AHES*, 8 (1936), 611 (Albitreccia).

    André Meynier later recalled that Bloch had asked him why he spent so much time discussing surfaces of erosion in a thesis devoted to the transformation of a poor agricultural region into a prosperous one. Claude Bataillon, "Table ronde imaginaire sur la géographie universitaire française, 1930–1940," *Hérodote*, 20 (1981), 140.

    Bloch made similar remarks in reviews of other works by geographers which were not strictly speaking regional monographs. See Marc Bloch, "Forêt et occupation du sol," *AHES*, 5 (1933), 495–496; Marc Bloch, "Du marais salant à l'herbage," *AHES*, 8 (1936), 605–606; Marc Bloch, review of *Islamisation de l'Afrique du Nord: les siècles obscurs du Magreb*, by E. F. Gautier, *BFLS*, 6 (1927–1928), 289–291.

32 Marc Bloch, "Une étude régionale: géographie ou histoire," *AHES*, 6 (1934), 81–85. Cf. Marc Bloch, "Un département et ses régions," *AHES*, 6 (1934), 527.

33 Marc Bloch, "Régions naturelles et groupes sociaux," p. 497; Marc Bloch, "Champs et villages," p. 473; Marc Bloch, "Un village breton", *AHES*, 8 (1936), 595–596; Marc Bloch, "Autour de la Loire," *AHES*, 10 (1938), 518–519; Marc Bloch, "Choses et gens du Sud-Ouest," *AHS*, 3 (1941), 109. Marc Bloch, review of *Etude comparée de la vie rurale pyrénéenne. . .*, by P. Birot, *AHS*, 3 (1941), 112.

34 Marc Bloch, "Une monographie géographique," p. 608; Marc Bloch, "Un 'cas' d'histoire agraire," *MHS*, 3 (1943), 94- 7. Here at least Bloch agreed with Febvre's early reviews.

35 Marc Bloch, "La France mediterranéenne," *RS*, 7 (1934), 269; Marc Bloch, "Points de vue sur le Limousin," *MHS*, 2 (1942), 77. In private, Bloch was more critical of Sion's book. See AN 318 MI 2, no. 441, letter of Lucien Febvre to Marc Bloch, n.d. (c. Mar. 1934).

36 Bloch, "Champs et villages," p. 471.

37 Ibid., p. 472.

38 Marc Bloch, "Une belle histoire humaine: nomadisme et vie sédentaire en Tunisie Orientale," *AHS*, 3 (1941), 164–165.

39 Cf. Marc Bloch, "Vieilles et nouvelles routes du Massif Central," *AHES*, 6 (1934), 185–186; Marc Bloch, "Dans la France du Centre," *AHES*, 8 (1936), 318 (re Meynier's *Géographie du Massif Central*); Marc Bloch, "Du marais salant à

l'herbage," p. 606. See also Bloch's lecture notes for three talks given in Brussels in 1932 on *le régime agraire* and his notes for a talk at Madrid the same year entitled "L'Enseignement historique d'un promenade dans un village français," both in AN AB^{xix} 3844.

40 Marc Bloch, "Champs et villages," pp. 468–469 (re Gerda Bernhard's *Das Nördliche Rheinhessen*); Bloch, "Paysages agraires du Nord," *AHS*, 3 (1941), 159–160 (re F. Mager's work on Schleswig).

41 Marc Bloch, "Régions naturelles et groupes sociaux," p. 506 (re F. Mager's work on Schleswig).

42 See, for example, Marc Bloch, "Du marais salant à l'herbage," pp. 605–606.

43 In some ways he was perhaps closest to his colleague Gabriel Le Bras who developed an historical perspective strongly influenced by sociology and was also sympathetic to geographical concerns. However, given Bloch's directorial position on the *Annales*, he was in a better position than Le Bras to express these views.

44 Marc Bloch, "Note sur deux ouvrages d'histoire et d'économie rurale," *RSH*, 27 (1913), 162.

45 Marc Bloch, "Les Paysages agraires," p. 277.

46 According to Toubert, Bloch's interdisciplinary approach here owed much to work in both Germany and England. In particular, he stresses the importance of the Seminar für Landesgeschichte and Siedlungskunde at Leipzig which was active during Bloch's stay at there in 1908. Pierre Toubert, "Préface," to Marc Bloch, *Les Caractères originaux de l'histoire rurale française* (Paris: Colin, 1988), pp. 7–13.

47 Marc Bloch, "Peuplement et régime agraire," *RSH*, 42 (1926), 97.

48 Ibid., p. 99.

49 Marc Bloch, "Les Edits sur les clôtures et les enquêtes agraires au XVIIIe siècle," *Bulletin de la Société d'Histoire Moderne*, 5th ser., 9 (1926), 213–216.

50 See for example, Marc Bloch, review of *L'Agriculture et les classes paysannes* by P. Raveau, *Revue Critique d'Histoire et de Littérature*, n.s. 93 (1926), 437. See also Bloch's notes on the works of Seebohm, Vinogradoff, and Maitland in AN AB^{xix} 3851 and the letter of Marc Bloch to Lucien Febvre, Sept. 11, 1929, re Maitland, AN 318 MI 1, no. 568–569.

51 Marc Bloch, "Une source peu connu d'histoire et de géographie rurale," *AG*, 35 (1926), 458. For Bloch on air photos see, for example, Marc Bloch, "Les Plans parcellaires: l'avion au service de l'histoire agraire en Angleterre," *AHES*, 2 (1930), 557–558. Due to the lack of documentary information, Bloch conducted his own study on the forms of plows. See AN AB^{xix} 3843 and AN 318 MI 1, no. 29, letter of Marc Bloch to Lucien Febvre, Sept. 27, 1928.

52 Marc Bloch, "Les Plans parcellaires," *AHES*, 1 (1929), 62.

53 Bloch wrote numerous articles on these maps. See ibid., pp. 60–73, 390–398; Marc Bloch, "Les Plans parcellaires: l'avion au service. . ."; Marc Bloch, "Une bonne nouvelle: l'enquête sur les plans cadatraux français," *AHES*, 4 (1932), 370–371; Marc Bloch, "Les Plans parcellaires: le travail qui se fait," *AHES*, 5 (1933), 374–375; Marc Bloch, "Les Plans parcellaires: les terroirs du Nord au lendemain de la Révolution," *AHES*, 7 (1935), 39–41; Marc Bloch, "Une nouvelle image de nos terroirs: la mise à jour du cadastre," *AHES*, 7 (1935), 156–159. See also his grant proposals arguing the need for their reproduction in AN AB^{xix} 3844. See also his lecture notes for a series of talks on *les régimes agraires* given at Brussels in 1932. AN AB^{xix} 3844.

54 Marc Bloch, "Le Problème des systèmes agraires envisagés particulièrement en France," in VIᵉ Congrès International des Sciences Historiques, *Résumés des Communications présentés au Congrès, Oslo 1928* (Le Comité Organisateur au Congrès, 1928), p. 264 (author's emphasis).

55 Ibid. In his description of the principal agrarian systems in France, Bloch identified four, instead of three. The category of fields which were generally open but where collective rights were less strong was replaced by "le système de la culture temporaire, sans soles réglées," and "le système des champs sans clôtures et sans obligations collectives (système 'provençal')." Although his categorization changed, he relied, as before, on the two criteria of enclosures and collective rights, reflecting his early work on *la vaine pâture.* Ibid., p. 265.

56 This series, started by Edv. Bull, was meant to stress the economic, social, and legal side of peasant life. In addition to encouraging work in Scandinavia, Bull hoped to promote work on the historical questions neglected by the geographical studies of European farming and pointed to France, Austria, and Sweden as places for which key historical questions still needed to be investigated. Frederik Stang, *Report on the Activities of the Institute for Comparative Research in Human Culture in the years 1927–1930* (Oslo: Instituttet for Sammenlignende Kulturforskning, 1930), serie C1–3, pp. 3–7, 19–22.

57 AN 318 MI 1, no. 53, letter of Marc Bloch to Lucien Febvre, Aug. 9, 1929.

58 See grant proposal written shortly before Bloch's course in AN ABˣⁱˣ 3844. Although Bloch emphasized the study of social constraints, he also appears to have been intrigued with the possibility of using this subject to study the relationships between the individual and the social. See undated "Introduction" in AN ABˣⁱˣ 3843.

59 Marc Bloch, *Les Caractères originaux de l'histoire rurale française* (Paris: Colin, 1964); Gaston Roupnel, *Histoire de la campagne française* (Paris: Plon, 1981); Roger Dion, *Essai sur la formation du paysage rural français* (Tours: Arrault, 1934).

  Dion's work was apparently inspired by Bloch's. As he wrote to Bloch regarding his thesis, where he first treated these questions: "Le vrai point de départ, en matière d'études agraires, c'est l'admirable livre que vous avez intitulé: *Les Caractères originaux de l'histoire rurale française.* J'ai reçu de lui l'impulsion première; sans lui mon chapitre sur l'économie rurale ancienne du Bassin parisien n'aurait pas existé." AN ABˣⁱˣ 3843, letter of Roger Dion to Marc Bloch, July 4, 1934. See also letters of Roger Dion to Marc Bloch in AN ABˣⁱˣ 3843: Jan. 11, 1934 & Oct. 27, 1934; cf. letter of Roger Dion to Albert Demangeon in DP Apr. 7, 1937; and Roger Dion, "A propos des paysages ruraux," *AG*, 45 (1936), 84–88.

60 Marc Bloch, *Les Caractères originaux*, pp. 35–36.

61 Ibid., p. 46.

62 Ibid., p. 64. Cf. Marcel Mauss, Contribution to a debate at the Première Semaine Internationale de Synthèse, *Civilisation: le mot et l'idée* (Paris: La Renaissance du Livre, 1930), 2nd fasc., pp. 74–129, 140–143. Reprinted in Marcel Mauss, *Oeuvres* (Paris: Editions de Minuit, 1969), II, pp. 456–479.

63 In contrast to his Oslo paper, Bloch noted an even greater role for collective constraints. In addition, temporary cultivation was dropped from his scheme on the grounds that it did not form a regular system.

64 Marc Bloch, *Les Caractères originaux*, pp. 36, 61.
65 Ibid., p. 64.
66 See, for example, Gaston Roupnel, *Histoire de la campagne française*, pp. 251–252; Roger Dion, *La Val du Loire* (Tours: Arrault & Co., 1934), pp. 467–470; J. L. M. Gulley, "The practice of historical geography: a study of the writings of Professor Roger Dion," *Tijdschrift voor Econ. en Soc. Geografie*, 52 (1961), 170. Cf. Marc Bloch, "L'Outillage rural," *Cahiers de Radio-France*, 9 (May 15, 1938), 442–447. There he stated clearly, "le choix de l'outil résultait d'un choix social." See also discussion above re inventions, p. 154.
67 Marc Bloch, *Les Caractères originaux*, pp. 56, 57.
68 For a contrasting interpretation, see Krzysztof Pomian, "L'Heure des *Annales*," in Pierre Nora, ed. *Les Lieux de Mémoire*, II, *La Nation*, I (Paris: Gallimard, 1986), pp. 395–397.
69 Dion, *Essai sur la formation*, pp. 3, 96ff., 124.
70 Cf. ibid., pp. 81, 93, 117; Marc Bloch, "Les Paysages agraires," pp. 256–260; Marc Bloch (pseud. "Fougères"), "Les Régimes agraires: quelques recherches convergentes," *AHS*, 3 (1941), 124. Demangeon and Sion sided with Bloch on this issue: Albert Demangeon, "Paysages ruraux," *AG*, 44 (1935), 538–539; Jules Sion, "Sur la structure agraire de la France Méditerranéenne," *Bulletin de la Société Languedocienne de Géographie*, 8 (1937), 14–15.
71 This difference in interpretation could be taken to reflect differences in both methodology and the object of study. As discussed by Alan Baker, Dion used what he called a retrospective method which sought to understand the present by the past whereas Bloch adopted *une méthode régressive* which sought to understand the past by the present, an approach used by both Seebohm and Maitland. Alan R. H. Baker, "A note on the retrogressive and retrospective approaches in historical geography," *Erdkunde*, 22 (1968), 244–245. For the earlier interpretations attacked by Baker, see J. L. M. Gulley, "The retrospective approach in historical geography," *Erdkunde*, 15 (1961), 306–309; H. C. Prince, "Historical geography in France," *Geographical Journal*, 124 (1958), 137–139. For a recent contrasting interpretation, see, Pomian, "L'Heure des *Annales*," p. 396.
72 Albert Demangeon, "L'Histoire rurale de la France," *AG*, 41 (1932), 239.
73 Jules Sion, "Une histoire agraire de la France," *RSH*, 3 (1932), 32.
74 Ibid., p. 31. See also, Jules Sion, "Note sur la notion de civilisations agraires," *Annales Sociologiques*, ser. E, 2 (1937), 74.
75 Albert Demangeon, "Une histoire de la campagne française," *AG*, 42 (1933), 410–415; Daniel Faucher, "Campagne française et campagnes méridionales à propos d'un livre récent," *Annales du Midi*, 45 (1933), 400–410; Demangeon, "Paysages ruraux," pp. 535–540; Sion, "Sur la structure agraire de la France Méditerranéenne," pp. 109–131; Sion, "Note sur la notion de civilisations agraires," pp. 71–79. See also, Faucher's discussion of Roupnel and Bloch in Daniel Faucher, "Polyculture ancienne et assolement biennal dans la France Méridionale," *Revue Géographique des Pyrénées et du Sud-Ouest*, 5 (1934), 241–255.
    Dion was very upset over the harsh criticism of his work by both Demangeon and Sion and so wrote formal rebuttals to each. Dion, "A propos des paysages ruraux," pp. 84–88; Roger Dion, "Sur la structure agraire de la France Méditerranéenne," *Bulletin de la Société Languedocienne de Géographie*, 2nd ser., 9

(1938), 1–7. See also Sion's answer, Jules Sion, "Observations," *Bulletin de la Société Languedocienne de Géographie*, 2nd ser., 9 (1938), 8–11 and Dion's letters to Demangeon, DP Oct. 17, 18, 19 (all probably 1935), Oct. 22, 1935, and Apr. 3 (probably 1936). For a more complimentary view see A. Gibert, "Sur le paysage rural français," *Etudes Rhodaniennes*, 11 (1935), 231–238.

76 EB letter of Marc Bloch to Lucien Febvre, Oct. 4, 1933.

77 The work in question was most probably Louis Halphen, *Les Barbares: des grandes invasions aux conquêtes turques du XIe siècle* (Paris: Alcan, 1926).

78 EB letter of Marc Bloch to Lucien Febvre, Oct. 4, 1933.

79 EB letter of Marc Bloch to Lucien Febvre, Feb. 20, 1936.

80 Marc Bloch, "Paysages agraires," pp. 256–260. Cf. Marc Bloch, "Les Servitudes collectives dans le Midi de la France," *AHS*, 3 (1941), 109–110. In later reviews of articles by Dion, Bloch noted that Dion had not significantly altered his earlier views, which Bloch had already criticized. Rather than repeat himself, Bloch praised Dion's efforts to demonstrate the limits of explanations which relied solely on geography. He praised Dion's "Souci de joindre à l'analyse, assurément indispensables, des facteurs géographiques, la vivifiante étude de réactions humaines, infiniment divers et dont les 'discordances' avec le milieu naturel sont souvent plus riches d'enseignements que la fameuse 'harmonie'." Marc Bloch, "Les Régimes agraires," pp. 118–124; Marc Bloch, "Le Forêt et les champs," *AHS*, 2 (1940), 165–166.

81 Marc Bloch, "Les Paysages agraires," p. 270.

82 Ibid., p. 273.

83 Marc Bloch, "Problèmes de structure agraire et de méthode," *MHS*, 2 (1942), 61–63. AN 318 MI 1, no. 318, letter of Marc Bloch to Lucien Febvre, Apr. 11, 1942. EB letter of Marc Bloch to Lucien Febvre, Oct. 4, 1933. See also his reviews: Marc Bloch, "Aux origines de notre société rurale," *MHS*, 2 (1942), 54; Marc Bloch, "Les Régimes agraires," p. 122; Marc Bloch, "Champs et villages," pp. 483–484; and his radio talk, Marc Bloch, "L'Outillage rurale."

84 "Séance du 12 Mars 1932: Marc Bloch, 'Le Problème des régimes agraires'," *Bulletin de l'Institut Français de Sociologie*, 2 (1932), 77–79.

85 Lucien Febvre, review of *Les Caractères originaux de l'histoire rurale française*, by Marc Bloch, *RH*, 169 (1932), 195. The works in question were Michel Augé-Laribé, *L'Evolution de la France agricole* (Paris: Colin, 1912); Henri Sée, *Les Classes rurales et le régime domanial en France au Moyen Age* (Paris: Gaird & Brière, 1901).

86 André Déléage, review of *Les Caractères*, *Annales du Midi*, 44 (1932), 219.

87 J. H. Clapham, review of *Les Caractères*, *English Historical Review*, 47 (1932), 655–657; R. H. Tawney, review of *Les Caractères*, *Economic History Review*, 4 (1933), 230–233.

88 As Heilbron pointed out, *sociologie* was not even listed in the statutes. Instead, the official goal of the society was to "rapprocher les spécialistes des diverses sciences sociales dont la réunion constitue la science de l'homme vivant en société." Johan Heilbron, "Les Métamorphoses du durkheimisme, 1920–1940," *RFS*, 26 (1985), 211.

89 "Séance du 12 Mars 1932," p. 49.

90 Ibid., p. 72.

91 Ibid., p. 86.

92 For a letter of Hubert to Bloch dated 4 July, 1928(?), see AN AB^{xix} 3843. Similarly, Bloch was skeptical of studies which labelled elements of early medieval European society as Germanic or Roman arguing that the issue was far more complex than that. Marc Bloch, "La Société du Haute Moyen Age et ses origines," pp. 403–420.

93 Bloch thus fits into Heilbron's group of those outside the discipline who provided "selective reinterpretations" of sociology. Heilbron, "Les Métamorphoses," p. 208.

94 For Bloch on *les brumes de préhistoire* see, for example, Marc Bloch, "La Sociologie et le passé du droit," *AHES*, 8 (1936), 458; Bloch, "Aux origines de notre société rurale," p. 46.

9 An expanding view: Marc Bloch's later projects

1 AN 318 MI 1, no. 189, letter of Marc Bloch to Lucien Febvre, Apr. 19, 1936. At Strasbourg, Bloch had taught a course dealing with *l'idée de l'Empire*. See Marc Bloch, "L'Empire et l'idée de l'Empire sous les Hohenstaufen," *Revue des Cours et Conférences*, 60 (1927–1928), 481–494, 577–589, 759–768.

2 See, for example, Marc Bloch, *L'Etrange défait* (Paris: Albin Michel, 1957), pp. 26–27; AN 318 MI 1, no. 269–274, 310–311, letters of Marc Bloch to Lucien Febvre, Apr. 6, 1940 and Aug. 17, 1941.

3 Cf. Henri Berr, "In memoriam," *RS*, 19 (1940–1945), 6.

4 HP letter of Marc Bloch to Henri Pirenne, Sept. 4, 1930; AN 318 MI 1, no. 53, letter of Marc Bloch to Lucien Febvre, Aug. 9, 1929. See also letter 39, Marc Bloch to Henri Pirenne, Jan. 6, 1927 in Lucien Febvre and Marc Bloch, *The Birth of Annales History: The Letters of Lucien Febvre and Marc Bloch to Henri Pirenne (1921–1935)*, eds. Bryce and Mary Lyon (Brussels: Académie Royale de Belgique, Commission Royale d'Histoire, 1991), letter 39, p. 89. There Bloch complains both of the enormity of this task and the need to leave aside his work on agrarian systems.

5 Letters of Marc Bloch to Henri Berr: Apr. 12, 1925; May 1, 1924; July 12, 1924, in Marc Bloch, *Ecrire La Société féodale: lettres à Henri Berr (1921–1943)*, ed. Jacqueline Pluet-Despatin (Paris: IMEC, 1992), letters 6, 1, 2, pp. 43, 28–29, 31–34.

6 For early plans of the series see the lists on the covers of the early volumes. At Bloch's request, he was also listed for volume no. 86, "La Révolution agricole." See Jacqueline Pluet-Desjardins, "Introduction," to Bloch, *Ecrire*, pp. 13–24 and the Preface to the same volume by Bronislaw Geremek, pp. 7–11.

7 Letter of Marc Bloch to Henri Berr, Feb. 8, 1933; letter of Marc Bloch to Lucien Febvre, Feb. 5, 1933; in *Ecrire*, letter 19, p. 76, and unnumbered letter, p. 71.

8 EB letters of Marc Bloch to Lucien Febvre, Oct. 9, 1933, piece of letter (detached), c. 1934; Feb. 14, 1942; June 22, 1942; AN 318 MI 1, no. 239, 310, 326–327, letters of Marc Bloch to Lucien Febvre, Dec. 5, 1938; Aug. 17, 1941; May 8, 1942. HP letters of Marc Bloch to Henri Pirenne, Mar. 11, 1927; Sept. 29, 1932. For early work on the economic history see manuscript in file marked "Les origines de l'économie européenne – cours 1927–1928" which includes an introduction to the book for Berr as well as a lengthy course, also intended, no doubt, as work toward the unfinished book. AN AB^{xix} 3823.

9 In addition to those discussed here, his posthumous article on the decline of slavery was also intended for that work. See above, pp. 119, 217 n50.

10 AN 318 MI 1, no. 143, letter of Marc Bloch to Lucien Febvre, Dec. 11, 1934. Bloch now described these articles as an "Esquisse de l'histoire du moulin d'eau dans un cadre d'histoire sociale." See AN 318 MI 1, no. 153, letter of Marc Bloch to Lucien Febvre, 13 Apr., 1935. For Febvre's criticism of this piece see AN 318 MI 3, no. 638, 750, letters of Lucien Febvre to Marc Bloch, n.d (both c. late 1935).

The articles in question were, Marc Bloch, "Avènement et conquêtes du moulin à eau," *AHES*, 7 (1935), 538–563; Marc Bloch, "Les 'Inventions' médiévales," *AHES*, 7 (1935), 634–643.

11 François Simiand, review of *La Picardie* and others, by Albert Demangeon and others, *AS*, 11 (1910), 729.

12 Marc Bloch, "Technique et évolution sociale: Réflexions d'un historien," *Europe*, 47 (1938), 23–32 as cited in Marc Bloch, *Mélanges historiques*, II (Paris: EHESS & Fleury, 1983), p. 838.

Contrary to Daniel Chirot, Bloch did not ignore the work of Lefebvre des Noëttes. In fact, Bloch had not only reviewed both editions of that work but had also published an answer to Lefebvre des Noëttes' rebuttal of his review. In addition to the article cited above see Marc Bloch, "Technique et évolution sociale," *RSH*, 41 (1926), 91–99; "Problèmes d'histoire des techniques," *AESC*, 4 (1932), 482–486; Commandant Lefebvre des Noëttes with an answer by Marc Bloch, "La Force animale et le rôle des inventions médiévales," *RSH*, 43 (1927), 83–91; Marc Bloch, "Les Inventions médiévales," *AHES*, 7 (1935), 639–643. Cf. Daniel Chirot, "The social and historical landscape of Marc Bloch," in Theda Skocpol, ed., *Vision and Method in Historical Sociology* (Cambridge, Cambridge University Press, 1984), pp. 22–46.

13 Cf. Bloch's file, "Les Origines de l'économie européenne – cours 1927–1928," in AN AB^xix 3823.

14 Marc Bloch, "Les Invasions: deux structures économiques," *AHS*, 7 (1945), 38. See also Marc Bloch, "Les Invasions: occupation du sol et peuplement," *AHS*, 8 (1945), 13–28. In his file "Les Origines de l'économie européenne" in AN AB^xix 3823, Bloch also spoke of the contrast between Germanic and Roman economies as one between civilizations. Other unpublished writings, apparently also intended for this work, include chapters on the social conditions of economic life and on the merchant class, along with others on the circulation of goods and economic conditions of exchange. Clearly, Bloch viewed this economic history as closely linked to social history.

15 See, for example, his discussion in Marc Bloch, "Les Invasions: deux structures économiques," pp. 33–46. Cf. Marc Bloch, "Classifications et choix des faits en histoire économique," *AHES*, 1 (1929), 255–256. There Bloch criticized the traditional division between agriculture, industry, and commerce, much as Simiand had done in 1903 (see above, p. 45), but Bloch also argued that greater attention should be paid to *l'évolution des techniques*.

In a later work, Simiand did address the question of innovations, suggesting that progress in inventions could be related to price cycles. Bloch criticized this theory as being too simplistic. Instead, he claimed that one ought to look "towards the internal structure of the society and towards the interactions of the diverse groups which form it." Marc Bloch, "Les Transformations des techniques comme problème de psychologie collective," *Journal de Psychologie Normale et Pathologique*, (Jan.–Mar., 1948) as cited in Marc Bloch, *Mélanges historiques*, II, p. 797.

16 Marc Bloch, "Economie-nature ou économie-argent: un pseudo-dilemme," *AHS*, 5 (1939), 7–16 as cited in Marc Bloch, *Mélanges historiques*, II, p. 877; see also, Marc Bloch, "Le Problème de l'or au Moyen Age," *AHES*, 5 (1933), 32. These two articles probably also resulted from his work toward Berr's book. Cf. François Simiand, "Méthode historique et science sociale," *RSH*, 6 (1903), 147.

17 See, for example, his course developed between 1940 and 1942. Marc Bloch, *Esquisse d'une histoire monétaire de l'Europe* (Paris: Colin, 1954).

18 Bloch's plans for his unwritten "Esquisse d'une histoire économique de la monnaie française" can be found in AN AB^{xix} 3824.

19 See François Dosse, *L'Histoire en miettes: des "Annales" à la "nouvelle histoire"* (Paris: Editions la Découverte, 1987), pp. 14–15. The works in question were François Simiand, *Recherches anciennes et nouvelles sur le mouvement général des prix du XVIe au XIXe siècle* (Paris: Donnat Montchrestien, 1932); Ernest Labrousse, *Esquisse du mouvement des prix et des revenues en France au XVIIIe siècle* (Paris: Dalloz, 1933); Henri Hauser, *Recherches et documents sur l'histoire des prix en France de 1500 à 1800* (Paris: Les Presses Modernes, 1936).

20 Marc Bloch, *La Société féodale* (Paris: Albin Michel, 1968), p. 16.

21 AN 318 MI 1, no. 264–265, letter of Marc Bloch to Lucien Febvre, Jan. 31, 1940.

22 Marc Bloch, *La Société féodale*, p. 15. For Bloch re social geography, see, for example, Marc Bloch, "Liberté et servitude personnelles du Moyen Age," *Anuario de Historia del Derecho Español*, (1933) as cited in *Mélanges*, I, p. 286; Marc Bloch, "Les Paysages agraires: essai de mise au point," *AHES*, 8 (1936), 270; Marc Bloch, *La Société féodale*, pp. 251, 431.

23 Marc Bloch, *La Société féodale*, p. 21.

24 Ibid., p. 214.

25 See also a lesson on "le problème de la féodalité" which Bloch gave in November, 1920 in AN AB^{xix} 3810. Bloch's interpretation here has been challenged recently by Elisabeth Magnou-Nortier. She argues that in his attempt to stress the master-slave relation, Bloch exaggerated the originality of feudal society and in so doing ignored important links both between the fief and the *bienfait* and between the "soi-disant nouveaux seigneurs et des familles aristocratiques." E. Magnou-Nortier, "Les 'Lois féodales' et la société d'après Montesquieu et Marc Bloch ou la seigneurie banale reconsidérée," *RH*, 586 (1993), 340, 330.

26 Marc Bloch, *La Société féodale*, p. 224.

27 See Henri Berr, "Introduction à une histoire universelle," *RSH*, 31 (1920), 19; Berr's introductions in which he did his best to draw links to other volumes; AN 318 MI 2, no. 502bis, letter of Lucien Febvre to Marc Bloch (c. June 1934); letters of Marc Bloch to Henri Berr, Feb 8. 1933 and Oct. 8, 1933 in Marc Bloch, *Ecrire*, letters 19 and 22, pp. 74–77, 82–83; Ferdinand Lot, review of *La Société féodale*, by Marc Bloch, *Journal des Savants* (Apr.–June, 1943), pp. 55–56.

28 At one stage Bloch intended to deal with such questions as the bourgeoisie and the clergy but later abandoned this idea in an effort to shorten his massive manuscript. See AN 318 MI 1, no. 325, letter of Marc Bloch to Lucien Febvre, May 8, 1942. See also what appears to be an early plan for such chapters on the back of a set of notes on Demangeon in AN AB^{xix} 3829.

29 Marc Bloch, *La Société féodale*, p. 13. For another reference by Bloch to stages of evolution see his 1939 address given at the "Deuxièmes Journées de Synthèse

Historique": Marc Bloch, "Sur les grandes invasions: quelques positions de prob-lèmes," *RS*, n.s. 19 (1945) 75.

30 At a recent Paris conference Bloch's treatment of feudalism was criticized for being Eurocentric. The conference proceedings are published in Hartmut Atsma and André Burguière, eds., *Marc Bloch aujourd'hui: histoire comparée et sciences sociales* (Paris: EHESS, 1990); see the chapters by Evelyn Patlagean, "Europe seigneurie, féodalité. Marc Bloch et les limites orientales d'un espace de comparai-son," pp. 279–298; Michel Cartier, "Les Historiens chinois, du marxism au com-paratisme. L'exemple du féodalisme," pp. 299–305; Lucette Valensi, "Retour d'Orient. De quelques usages du compartisme," pp. 307–316.

31 See Marcel Granet, *La Féodalité chinoise* (Paris: Imago, 1981). Cf. Marc Bloch, "Les Féodalités: une enquête comparative," *AHES*, 8 (1936), 591.

32 Cf. Louis Halphen, *Les Barbares: des grandes invasions aux conquêtes turques du XIe siècle* (Paris: Alcan, 1926).

33 Marc Bloch, *La Société féodale*, p. 23.

34 EB letter of Marc Bloch to Lucien Febvre, Oct. 4, 1933.

35 See, for example, Marc Bloch, "Les Plans parcellaires," *AHES*, 1 (1929), 62. Cf. Bernard Bailyn, "Braudel's geohistory – a reconsideration," *Journal of Economic History*, 11 (1951), 280, where Bailyn contrasts Bloch's approach with that of Braudel.

36 This is possibly because of the time of its publication at the start of the war, just after the last number of the *Annales Sociologiques* on "sociologique juridique et morale," and long before the publication of the third series of the *Année Sociologique*.

37 Lucien Febvre, "La Société féodale," *AHS*, 2 (1940), 39–43; Lucien Febvre, "La Société féodale," *AHS*, 3 (1941), 125–130; Ch.-Edmund Perrin, "La Société féodale: à propos d'un ouvrage récent," *RH*, 194 (1944), 130; Lot, review of *La Société féodale*, p. 56; Paul Ourliac, review of *La Société féodale, Bibliothèque de l'Ecole des Chartes*, 11 (1941), 218–223; F. M. Powicke, review of *La Société féodale, English Historical Review*, 55 (1940), 449–451. See also William A. Morris, *AHR*, 45 (1940), 855–856; 46 (1941), 617–618, who complained that the plan of the volume "does not admit of precise documentation"; F. L. Ganshof, *Revue Belge de Philologie et d'Histoire*, 20: 1–2 (1941), 183–193, who, in a largely favorable review, pointed to the need for a more rigorous juridical analysis.

  For more recent assessments criticizing the work for being overly sociological, see Bryce Lyon, "The feudalism of Marc Bloch," in Bryce Lyon, *Studies of West European Medieval Institutions* (London: Variorum Reprints, 1978), pp. 275–283; Elisabeth Magnou-Nortier, "Les 'Lois féodales'," pp. 321–360. Magnou-Nortier claims that Bloch's book was seductive but unfounded; Bloch, she writes, exagger-ated the difference between feudal society and that which preceded it. In opposi-tion to Bloch, she makes much stronger links to the political realm and stresses legal classifications.

  Robert Fossier attributes the lack of a strong critical response to *La Société féodale* both to the war and anti-Semitism and also to its innovative character: "l'ouvrage dérangeait les certitudes, bouleversait les schémas traditionnels." Robert Fossier, "Préface," to Marc Bloch, *La Société féodale* (Paris: Albin Michel, 1989), p. ii.

38 EB Dossier Gallimard, G II 2 (lettres de l'éditeur), Gaston Gallimard to Marc Bloch, Nov. 2, 1934. Bloch had been approached in 1933 to contribute a book on *l'homme et les champs*" to Deffontaine's series, "La Géographie Humaine" (also published by Gallimard), but he declined. EB letter of Marc Bloch to Lucien Febvre, Nov. 22, 1933.

39 EB Dossier Gallimard, G II 10 (Seigneurie française et manoir anglais), letter which appears to be one of Marc Bloch to Eileen Power, Mar. 2, 1935.

40 See "Proposition Gallimard," which appears to be Bloch's preparatory notes for an initial meeting in EB Dossier Gallimard, G II 4. On Bloch's desire for *souplesse* see also letter of Marc Bloch to Henri Pirenne, May 29, 1935, in Lyon and Lyon, eds., *The Birth of Annales History,* letter 82, p. 171.

41 EB Dossier Gallimard, G II 2 (lettres de l'éditeur), letter from Gaston Gallimard to Marc Bloch, Nov. 23, 1934.

42 EB letter of Marc Bloch to Lucien Febvre, Nov. 21, 1934. Bloch also wrote Pirenne about his desire to "procéder par problèmes plutôt que d'offrir une carte d'échantillons de tous les 'paysans' de la terre." Letter of Marc Bloch to Henri Pirenne, May 29, 1935, in Lyon and Lyon eds., *The Birth of Annales History,* letter 82, p. 171.

43 EB Dossier Gallimard, G I (lettres à l'éditeur), letter of Marc Bloch to Gaston Gallimard, Mar. 9, 1935.

44 See Jules Sion, "Géographie et ethnologie," *AG,* 46 (1937), 464.

45 Ibid.

46 Among those initially approached by Bloch were four of his fellow social historians at Strasbourg (Piganiol, Le Bras, Lefebvre, Febvre). Piganiol and Le Bras both contributed to later series of the *Année Sociologique* and Lefebvre and Febvre, while not quite so close to the Durkheimians, were very familiar with their work. Bloch also contacted three ethnographers: André Varagnac (who had studied with Henri Hubert), Jacques Soustelle (who was closely linked with Mauss – see: AN 318 MI 2, no. 592, letter from Lucien Febvre to Marc Bloch, n.d. – c. Mar. 1935), and Henri Labouret an early collaborator of the *Annales.* Other initial contacts included Sion (the most Durkheimian of the Vidalians), a few foreign historians (K. Asakawa, M. Olsen, Lis Jacobsen), and André Siegfried. Later names added while Bloch was still fairly actively involved included three social historians (Jean Gagé, Robert Boutruche, and Lucie Varga), the geographer J. Célérier (a friend of Bloch's also interested in social history), and Jean-Paul Hütter (a student of Hauser's). The publisher also proposed the names of Roger Dion and P. Pascal. During the war years, new names included five geographers: Aimé Perpillou, Paul Marrès, Jules Blache (all three proposed by the acting director André Déléage), Pierre Gourou (whom Bloch contacted directly), and Jacques Weuleuresse.

    In addition to their links to Durkheim, the presence of the ethnographers probably also reflected the provincial nature of French geography during this period when few geographers studied areas outside of France. See Numa Broc, "Homo geographicus: radioscopie des géographes français de l'entre-deux guerres (1918–1939)," *AG,* 102 (1993), 231. See also Jules Sion, "Géographie et ethnologie," *AG,* 46 (1937), 464.

47 EB Dossier Gallimard, G II 5 (Hütter), letter of Jean-Paul Hütter to Marc Bloch, Feb. 12, 1936.

48 EB Dossier Gallimard, G II 6 (Sion), letter of Jules Sion to Marc Bloch, Mar. 20, 1935.
49 Jacobsen, for example, could restrict her study on Scandinavia to the Middle Ages and Asakawa could focus on the pre-modern period in Japan. Bloch did however encourage the latter to include a few pages on the "present capitalist age" as Asakawa put it and tried to get Sion to give greater historical treatment than *les coupes de sondes* which he had initially proposed. EB Dossier Gallimard: G II 14 (Asakawa), letter of Asakawa to Marc Bloch, Jan. 21, 1935; G II 6 (Sion) letters of Jules Sion to Marc Bloch, Feb. 5, 1935 and Mar. 20, 1935, G II 13 (Scandinavie), letter of Marc Bloch to Lis Jacobsen, May 13, 1935.
50 EB Dossier Gallimard: G II 12 (Labouret), letter of Marc Bloch to Henri Labouret, Jan. 18, 1934; G II 13 (Scandinavie), letters of Marc Bloch to Lis Jacobsen, May 13, 1935 and Feb. 2, 1935; G II 6 (Sion), letter of Jules Sion to Marc Bloch, June 18, 1938. See also letters of Jules Sion to Marc Bloch, Mar. 12 and Mar. 20, 1935, in same file.
      The contrast between Sion and Bloch was also illustrated by Bloch's contribution to the section on food, edited by Sion, for the *Encyclopédie Française*. See Sion's proposals and letters to Bloch in AN AB[xix] 3844. Cf. Marc Bloch, "Dans la France ancienne," *Encyclopédie Française* (Paris: Société Nouvelle de l'Encyclopédie, 1955), XIV, 14.40.2 – 14.40.3; Marc Bloch, "Les Aliments du Français," *Encyclopédie Française*, XIV, 14.42.7 – 14.42.10.
51 EB Dossier Gallimard (Hütter), letter of J.-P. Hütter to Marc Bloch, Feb. 21, 1936.
52 EB Dossier Gallimard, G II 13 (Scandinavie), letter of Marc Bloch to Jacobsen, May 13, 1935.
53 EB Dossier Gallimard, G II 12 (Labouret), letter of Marc Bloch to Labouret, Jan. 18, 1934; G II 13 (Scandinavie) letter of Marc Bloch to Jacobsen, Feb. 2, 1935.
54 Marc Bloch, "Note de l'éditeur" to Henri Labouret, *Paysans d'Afrique Orientale* (Paris: Gallimard, 1941), p. 5.
55 Ibid., p. 6.
56 Another volume, not included in Bloch's original plans, was published in his lifetime, *Le Village et le paysan de France* by Alfred Dauzat. This had been accepted during the war without Bloch's knowledge. Dauzat relied heavily on geographical explanations and dwelled on material appearances without always seeking what Bloch would see as the underlying social reality. He gave little sense of the complexity and diversity of social life, spoke instead of "l'âme paysan", and relied on stereotypes and popularized accounts of folklore. Understandably, Bloch was not at all pleased with the result. Alfred Dauzat, *Le Village et le paysan de France* (Paris: Gallimard, 1941); AN 318 MI 1, no. 326, letter of Marc Bloch to Lucien Febvre, May 8, 1942.
57 EB Dossier Gallimard, G II 3 (La Collection en Général), letter of Henri Labouret to Marc Bloch, Mar. 13, 1939.
58 EB Dossier Gallimard, G II 12 (Labouret), letter of Henri Labouret to Marc Bloch, Aug. 18, 1938.
59 Marc Bloch, *Seigneurie française et manoir anglais* (Paris: Colin, 1967), pp. 14, 17, 69.
60 Marc Bloch, "La Genèse de la seigneurie: idée d'une recherche comparée," *AHES*, 9 (1937), 225–227. Cf. Marc Bloch, "Seigneurie française et manoir anglais:

quelques problèmes d'histoire comparée" (talk given at the London School of Economics) in AN AB[xix] 3834.

61 AN 318 MI 1, no. 252, 327, letter of Marc Bloch to Lucien Febvre, Oct. 8, 1939; Marc Bloch, "Réflexions pour le lecteur curieux de méthode," *Rivista di Storia della Storiografia Moderna*, 9:2–3 (1988), 170–173. For first chapter, see AB[xix] 3831, IIb 18.

62 AN 318 MI 1, no. 327, letter of Marc Bloch to Lucien Febvre, May 8, 1942. EB letter of Marc Bloch to Lucien Febvre, June 22, 1942. Bloch, in fact did begin to assemble material for a study of the First Reich, see AN AB[xix] 3809. According to Etienne Bloch, other projects included a whodunit entitled "un meurtre de province" and a "manuel pour l'étude de l'histoire rurale française." The latter was to be a collective manual, possibly intended for the series "Le Paysan et la terre," to which Bloch intended to contribute a number of chapters. Etienne Bloch, "Marc Bloch. Souvenirs et réflexions d'un fils sur son père," in Atsma and Burguière, eds., *Marc Bloch aujourd'hui,* p. 35.

63 AN 318 MI 1, no. 349, letter of Marc Bloch to Lucien Febvre, Oct. 17, 1942.

64 Marc Bloch, *Apologie pour l'histoire ou métier d'historien* (Paris: Colin, 1974), p. 30. The published versions of this work unfortunately contain a number of errors and omissions. See Massimo Mastrogregori, "Nota sul testo dell *Apologie pour l'histoire* di Marc Bloch," *Rivista di Storia della Storiografia Moderna*, 7:3 (1986), 5–32; Massimo Mastrogregori, "Le manuscrit interrompu: *Métier d'Historien* de Marc Bloch," *Annales ESC*, 44:1 (Jan.–Feb., 1989), 147–159. Here, I have used the 1974 edition amended by a list of corrections kindly sent to me by Marleen Wessel.

65 AN 318 MI 1, no. 310, letter of Marc Bloch to Lucien Febvre, Aug. 17, 1941. As early as 1934, Bloch had considered bringing together a number of his methodological articles into book form to be published by Gallimard and in 1939 enjoyed writing the methodological section for his incomplete manuscript "Histoire de la société française dans le cadre de la civilisation européenne," later published as Marc Bloch, "Réflexions pour le lecteur" EB Dossier Gallimard, G II 2, letter of Gaston Gallimard to Marc Bloch, Nov. 23, 1934; AN 318 MI 1, no. 252, letter of Marc Bloch to Lucien Febvre, Oct. 8, 1939.

66 AN 318 MI 1, no. 348, letter of Marc Bloch to Lucien Febvre, Oct. 17, 1942. Febvre had earlier made a similar objection to Bloch's review of Paul Harsin's *Comment on écrit l'histoire.* AN 318 MI 3, no. 711, letter of Lucien Febvre to Marc Bloch, n.d. (c. late 1935).

67 AN 318 MI 1, no. 334, letter of Marc Bloch to Lucien Febvre, Aug. 17, 1943.

68 "Séance du 30 Mai, 1907: les conditions pratiques de la recherche des causes dans le travail historique," *Bulletin de la Société Française de Philosophie*, 7 (1907), 304 ; Marc Bloch, *Apologie pour l'histoire*, pp. 28–29. Bloch had earlier criticized Langlois for suggesting that one could convey a good sense of history simply by presenting a carefully chosen collection of texts. By contrast, Bloch argued that active historical reconstruction was essential, a task which required imagination. Marc Bloch, "Quelques contributions à l'histoire religieuse du Moyen Age," *RSH,* 47 (1929), 95–97.

69 Marc Bloch, *Apologie pour l'histoire*, p. 27.

70 Ibid.

71 Ibid., p. 29.

72 On relativity, Bloch wrote for example: "Il y a longtemps, en effet, que nos grands ainés, un Michelet, un Fustel de Coulanges nous avaient appris à le reconnaître: l'objet de l'histoire est par nature l'homme. Disons mieux 'les hommes.' Plutôt que le singulier, favorable à l'abstraction, le pluriel, qui est le mode grammatical de la relativité, convient à une science du divers." Marc Bloch, *Apologie pour l'histoire*, pp. 34–35, see also, pp. 36, 157; cf. Marc Bloch, "Technique et évolution sociale: reflexions d'un historien," as cited in Marc Bloch, *Mélanges historiques*, II, p. 832.

73 Charles-Victor Langlois and Charles Seignobos, *Introduction aux études historiques* (Paris: Hachette, 1898), pp. 188, 205, 235. Seignobos also labeled as voluntary precisely those phenomena such as language which sociologists would see as best illustrating a social fact. Ibid., pp. 208–209.

74 EB letter of Marc Bloch to Lucien Febvre, Dec. 13, 1935. For examples of Bloch's willingness to examine the social in *Apologie* see Marc Bloch, *Apologie pour l'histoire*, pp. 58, 70, 95, 99, 101, 158. On the individual and collective in Bloch's *Apologie pour l'histoire* see interesting discussion in Marleen Wessel "Woord Vooraf," pp. 21–23 – her introduction to her Dutch translation: Marc Bloch, *Pleidooi voor de Geschiedenis of Geshiedenis als Ambacht* (Nijmegen: SUN, 1989).

75 Marc Bloch, *Apologie pour l'histoire*, pp. 158, 91–92; cf. "Séance du 30 Mai 1907: les conditions pratiques de la recherche des causes dans le travail historique," p. 306. See also, François Simiand, "Méthode historique et science sociale," *RSH*, 6 (1903), 18–19.

76 Marc Bloch, *Apologie pour l'histoire*, p. 31.

77 Ibid., p. 120. Bloch apparently planned a more thorough study of the relative roles of the individual and the social since in an outline described by Febvre he had included a later section described as follows: "Le problème de l'individu et de sa valeur différentielle. Accessoirement, les épochs documentairement sans individus – l'histoire est-elle seulement une science des hommes en société? L'histoire-masse et les élites." Unfortunately this was one of the sections left unwritten in the incomplete manuscript. See Lucien Febvre, "Comment se présentaient les manuscrits de 'Métier d'historien,'" appendix to Bloch, *Apologie pour l'histoire*, p. 163.

78 Marc Bloch, *Apologie pour l'histoire*, p. 147. In addition, he attacked the terms "Middle Ages" and "Renaissance" which, he claimed, imposed a division based on what was thought to be a progress of the spirit on all of history.

79 Marc Bloch, *Apologie pour l'histoire*, p. 151.

80 Ibid., p. 153. Bloch also used Bergson's term *élan vital.* See Marc Bloch, "Régions naturelles et groupes sociaux," *AHES*, 4 (1922), 506.

81 Marc Bloch, *Apologie pour l'histoire*, p. 143.

82 Ibid., p. 139.

83 Ibid., p. 131. Cf. Robert Boutruche, "Marc Bloch vu par ses élèves," in *Mémorial des années 1939–1945* (Paris: Belles Lettres, 1947), Publications de la Faculté des Lettres de l'Université de Strasbourg, fasc. 103, p. 200.

84 Langlois and Seignobos, *Introduction aux études historiques*, pp. 249–250; Marc Bloch, *Apologie pour l'histoire*, p. 124.

85 AN 318 MI 1, no. 217–218, letter of Marc Bloch to Lucien Febvre, June 22, 1938.

86 Cf. Simiand, "Méthode historique et science sociale," pp. 141–143, 146.

87 Marc Bloch, *Apologie pour l'histoire*, pp. 125–126.

88 Ibid., pp. 159–160.

89 Ibid. p. 65.
90 Ibid., p. 35.
91 Ibid., pp. 44, 49.
92 Ibid., p. 126.

10 Towards a reworking of the historiography of Marc Bloch

1 Although some comparisons have been made to Durkheim, frequently Bloch's comparative approach has been examined in light of other traditions. See, for example, Raftis re humanism, Sewell and hypothesis testing, Walker and model building, the Hills and linguistics: J. Ambrose Raftis, "Marc Bloch's comparative method and the rural history of medieval England," *Medieval Studies*, 24 (1962), 349–368; William H. Sewell, Jr., "Marc Bloch and the logic of comparative history," *History and Theory*, 6 (1967), 208–218; Lawrence Walker, review of *Feudal Society* by Marc Bloch, *History and Theory*, 3 (1963), 251–253; Alette Olin Hill and Boyd H. Hill, Jr., "Marc Bloch and comparative history," *American Historical Review*, 85 (1980), 828–846. See also comments on the Euro-centric character of Bloch's comparisons at a 1986 conference commemorating Bloch by Patlagean, Cartier, and Valensi published in Hartmut Atsma and André Burguière, eds., *Marc Bloch aujourd'hui: histoire comparée et sciences sociales* (Paris: EHESS, 1990): Evelyn Patlagean, "Europe seigneurie, féodalité. Marc Bloch et les limites orientales d'un espace de comparaison," pp. 279–298; Michel Cartier, "Les Historiens chinois, du marxisme au comparatisme. L'exemple du féodalisme," pp. 299–305; Lucette Valensi, "Retour d'Orient. De quelques usages du comparatisme," pp. 307–316.
2 For examples of authors arguing that Bloch's method was essentially fixed as early as 1914, see, Lucien Febvre, preface to Marc Bloch, "Critique historique et critique du témoignage," *AESC*, 5 (1950), 1; Carole Fink, *Marc Bloch: A Life in History* (Cambridge: Cambridge University Press, 1989), p. 50. For a contrasting interpretation see Carlo Ginsburg, "A proposito della raccolta dei sagistorici di Marc Bloch," *Studi Medievali*, 6 (1965), 335–353.
3 Theda S. Skocpol, "Sociology's historical imagination," pp. 1–21, and "Emerging agendas and recurrent strategies in historical sociology," pp. 356–391 in Theda Skocpol, ed., *Vision and Method in Historical Sociology* (Cambridge University Press, 1984); Daniel Chirot, "The social and historical landscape of Marc Bloch," in Skocpol, ed., *Vision and Method,* pp. 22–46; Denis Smith, *The Rise of Historical Sociology* (Cambridge: Polity Press, 1991). See especially, pp. 156, 166.
    For endorsements of Bloch's approach by sociologists see, for example, R. Di Donato, "L'Anthropologie historique de Louis Gernet," *AESC*, 37 (1982), 987 (on Louis Gernet and Bloch); J. Stengers, "Marc Bloch et l'histoire," *AESC*, 8 (1953), 331; G. Noiriel, "Pour une approche subjectiviste du social," *AESC*, 44 (1989), 1444.
4 Hugh D. Clout, ed., "The practice of historical geography in France," in Hugh D. Clout, ed., *Themes in the Historical Geography of France* (London: Academic Press, 1977), pp. 1–19; Alan R. H. Baker, "Reflections on the relations of historical geography and the Annales school of history," in Alan R. H. Baker and Derek Gregory, ed., *Explorations in Historical Geography* (Cambridge: Cambridge University

Press, 1984), pp. 1–27; Paul Claval, "The historical dimension of French geography," *Journal of Historical Geography*, 10 (1984), 229–245.

5 See, for example, Henri Baulig, "Marc Bloch géographe," *AHS*, 8 (1945), 5–12; Etienne Julliard, "Aux frontiers de l'histoire et de géographie," *RH*, 215 (1956), p. 267–273.

6 On the development of historical geography in France, see discussion in Xavier de Planhol, "Historical Geography in France," in Alan R.H. Baker, ed., *Progress in Historical Geography* (Newton Abbot: David and Charles, 1972), pp. 29–44; Xavier de Planhol, "Structures universitaires et problématique scientifiques: la géographie française," in *La Pensée géographique française contemporaine* (Saint-Brieuc: Presses Universitaires de Bretagne, 1972), pp. 155–165; Robin Butlin, "A short chapter in French historical geography: the *Bulletin du Comité Français de Géographie Historique,*" *Journal of Historical Geography*, 16 (1990), 438–445; Robin Butlin, *Historical Geography: through the gates of time and space* (London: Arnold, 1993), pp. 30–31.

For a few examples of agrarian structure debate within geography drawing on Bloch's work, see L. Champier, "La Recherche française en matière d'histoire et de géographie depuis un quart de siècle," *Revue Géographique de Lyon*, 31 (1956), 319–327; Etienne Julliard, et al., "Structures agraires et paysages ruraux," *Annales de l'Est*, mémoire 17 (1957); Paul Claval, "L'Etude de l'habitat rural et les congrès internationaux de géographie," in *Geography through a Century of International Congresses* (IGU, Commission on the History of Geographical Thought, 1972), pp. 131–145.

7 Georges Duby, "Preface," to Marc Bloch, *Apologie pour l'histoire ou métier d'historien* (Paris: Colin, 1974), p. 11. For a good example of the use of Dion as a precedent by French historical geographers, see de Planhol, "Historical geography in France." A more recent example which refers to Bloch but still focusses on Dion is Marcel Roncayolo, "Histoire et géographie: les fondements d'une complémentarité," *AESC*, 44 (1989), 1427–1434.

8 See discussion in J. L. M. Gulley, "The practice of historical geography: a study of the writings of Professor Roger Dion," *Tijdschrift voor Econ. en Soc. Geografie*, 52 (1961), 169–175 and in Hugh Prince, "Regional contrasts in agrarian structure," in Hugh D. Clout, ed., *Themes in the Historical Geography of France* (London: Academic Press, 1977), pp. 129–184.

9 For Darby's recollections on the development of the field, see H.C. Darby," Some reflections on historical geography," *Historical Geography*, 9:1, 2 (1979), 9–13. For the proceedings of the joint meeting of the Historical and Geographical Associations see "What is historical geography?" *Geography*, 17 (1932), pp. 39–45.

10 Cole Harris, "Theory and synthesis in historical geography," *The Canadian Geographer*, 15 (1971), 163; Andrew H. Clark, "The whole is greater than the sum of its parts," in Donald R. Deskins, Jr., et al., *Geographic Humanism, Analysis and Social Action* (Ann Arbor: Michigan Geographical Publication no. 17, 1977), p. 13; Alan R. H. Baker, "Rethinking historical geography," in Alan R. H. Baker, ed., *Progress in Historical Geography* (Newton Abbott: David and Charles, 1972), p. 28; H. C. Darby, "Historical geography," in H. P. R. Finberg, ed., *Approaches to History*, (London: Routledge and Kegan Paul, 1962), pp. 135, 44, 148; Hugh Prince, "Real, imagined and abstract worlds of the past," *Progress in Geography*, 3

(1969), 7, 17, 18, 24, 27, 40; Alan R. H. Baker, "On the historical geography of France," review of *Themes in the Historical Geography of France* by Hugh D. Clout, *Journal of Historical Geography*, 6 (1980), 75.

11 For a good early example, see Charles-Edmund Perrin, "L'Oeuvre historique de Marc Bloch," *RH*, 199 (1948), 183. More recent examples include Peter Burke, *The French Historical Revolution: The Annales School, 1929–1989* (Cambridge: Polity Press, 1990), p. 15; Smith, *The Rise of Historical Sociology*, pp. 42–43. See also Davies re Bloch's "essentially sociological approach" and his criticism of geography, R. R. Davies, "Marc Bloch," *History*, 52 (1967), 278, 275; and Dosse re Bloch's use of sociological categories, François Dosse, *L'Histoire en miettes: des "Annales" à la "nouvelle histoire"* (Paris: Editions la Découverte, 1987), p. 52.

12 See, for example, Davies, "Marc Bloch," p. 275; Burke, *The French Historical Revolution*, p. 13.

13 Chirot, "The social and historical landscape," p. 39.

14 See, for example, Skocpol, "Sociology's historical imagination," p. 31.

15 Examples include Chirot, "The social and historical landscape"; Skocpol, "Emerging agendas" and Sewell, "Marc Bloch and the logic."

16 See R. Colbert Rhodes, "Emile Durkheim and the historical thought of Marc Bloch," *Theory and Society*, 5 (1978), 45–73; and on comparison, Jelle C. Riemersma, "Introduction to Marc Bloch," in Frederic C. Lane and Jelle C. Riemersma, *Enterprise and Secular Change: Readings in Economic History* (Homewood, IL: Irwin, 1953), pp. 493. For an example drawing parallels without distinguishing the approaches, see Fink, *Marc Bloch: A Life in History* (Cambridge University Press, 1989), pp. 42, 118. For an exception that goes farther than most in distinguishing Bloch's comparative method from that of Durkheim, see Massimo Mastrogregori, *Il Genio dello Storico: Le Considerazioni sulla Storia di Marc Bloch e Lucien Febvre e la Tradizione Methodologica Francese* (Rome: Edizioni Scientifiche Italiano, 1987), pp. 151–152.

17 See, for example, Burke, *The French Historical Revolution*, p. 19; Peter Burke, *Sociology and History* (London: Allen and Unwin, 1980), p. 77.

18 For a start see Carlo Ginsburg, "A proposito della raccolta dei sagistorici di Marc Bloch," *Studi Medievali*, 6 (1965), 335–353.

19 See, for example, Smith, p. 42; Baulig, pp. 5–12; Krzysztof Pomian, "L'Heure des Annales," in Pierre Nora, *Les Lieux de mémoire*, vol. II, part 1 (Paris: Gallimard, 1986), p. 392; Hervé Coutau-Bégarie, *Le Phénomène nouvelle histoire: grandeur et décadence de l'école des Annales* (Paris: Economica, 1989), p. 57.

20 Fink, *Marc Bloch,* pp. 46, 154; Burke, *The French Historical Revolution*, p. 15.

21 See for example, Chirot, "Social and historical landscape of Marc Bloch," pp. 32–33.

22 For an interesting start in this direction see Mastrogregori, *Il Genio*, p. 154 on Bloch's depiction of a "living and human reality." See also Lucien Febvre on Bloch and the "taste of the real": Lucien Febvre, "Marc Bloch et Strasbourg: souvenirs d'une grande histoire," in *Mémorial des années 1939–1945* (Paris: Les Belles Lettres, 1947), Publications de la Faculté des Lettres de l'Université de Strasbourg, fasc. 103, p. 176.

23 Here, choosing to emphasize the institutional context of the Université, I have only touched upon such issues as nation-building, regionalism, the relationships

between science and society, and Bloch's strong republican values. Two very useful accounts of the need for broadly based contextual work within the field of the history of geography are David N. Livingstone, *The Geographical Tradition: Episodes in the History of a Contested Enterprise* (Oxford: Blackwell, 1992), esp. pp. 1–31; Henry Aay, "Textbook chronicles: disciplinary history and the growth of geographic knowledge," in Brian Blouet, ed., *The Origins of Academic Geography in the United States* (Hamden, Conn.: Archon, 1981), pp. 291–301.

24 Bryce Lyon, "The feudalism of Marc Bloch," in Bryce Lyon, *Studies of West European Medieval Institutions* (London: Variorum Reprints, 1978), pp. 275–283 – originally read as a paper at the 25th Vlaams Filiogencongres, held at Antwerp, Apr. 17–19, 1963 and published in *Tijdschrift voor Geschiedenis*, 76 (1963).

25 By contrast some others have implied a less critical adoption of the geographers' regional approach by Bloch: Davies, "Marc Bloch," p. 275; André Burguière, "La Notion de 'mentalités' chez Marc Bloch et Lucien Febvre: deux conceptions, deux filiations," *RS*, 3rd ser., 111–112 (1983), p. 339.

26 Marc Bloch, "Sur les grandes invasions: quelques positions de problèmes," *RS*, n.s. 19 (1945), 75 (originally a paper given at the Deuxièmes Journées de Synthèse Historique, May 31–June 9, 1939).

27 See discussion contrasting Bloch's form of explanation and that more typical of historical sociology in Olivier Zunz, "Toward a dialogue with historical sociology," *Social Science History*, 2:1 (1987), 36–37; André Burguière, "The *Annales*: continuity and discontinuity," unpublished paper cited in above.

28 The term is from the essay in Clifford Geertz, *Local Knowledge* (New York: Basic Books, 1983) pp. 19–35. See also discussion in Anthony Giddens, *Profiles and Critiques in Social Theory* (Berkeley: University of California Press, 1982), p. 6. For France see the editorial call for papers: "Histoire et sciences sociales. Un tournant critique," *AESC*, 43 (1988), 291–293. See also the special issue which resulted, *AESC*, 44 (1989).

29 Anthony Giddens, *Central Problems in Social Theory: Action, Structure and Contradiction in Social Analysis* (Berkeley and Los Angeles: University of California Press, 1979); Anthony Giddens, *Profiles and Critiques in Social Theory* (Berkeley and Los Angeles: University of California Press, 1983); Anthony Giddens, *The Constitution of Society* (Berkeley and Los Angeles: University of California Press, 1984); Allan Pred, "Power, everyday practice and the discipline of human geography," in *Space and Time: Essays Dedicated to T. Hägerstrand* (Lund: Glerrups, 1981); Allan Pred, "Place as historically contingent process: structuration and the time-geography of becoming places," *Annals of the Association of American Geographers*, 74 (1984), 279–297; Derek Gregory, *Social Theory and Spatial Structure* (London: Macmillan, 1982); Derek Gregory and John Urry, eds., *Social Relations and Spatial Structures* (London: Macmillan, 1985); Derek Gregory, "Space, time and politics in social theory; an interview with Anthony Giddens," *Environment and Planning D: Society and Space*, 1984 (2), 123–132; Derek Gregory, "Presences and absences: time-space relations and structuration theory," in David Held and John B. Thompson, eds., *Social Theory of Modern Societies: Anthony Giddens and his Critics* (Cambridge University Press, 1989), pp. 185–214; Anthony Giddens, "A reply to my critics," in Held and Thompson, *Social Theory*, pp. 275–291.

Closely related to the structuration debate are the attempts to promote a "reconstructed regional geography" drawing heavily on developments within social history and social science. See Nigel Thrift, "On the determination of social action in space and time," *Environment and Planning D: Society and Space*, 1 (1983), 23–57; Mary Beth Pudup, "Arguments within regional geography," *Progress in Human Geography*, 12 (1988), 369–390. Also of related interest is the more broadly based effort to promote a "new" cultural geography, which draws on a wide range of contemporary social and cultural theory including structuration theory. See Peter Jackson, *Maps of Meaning* (London: Unwin Hyman, 1989).

30 Although some efforts have been made to modify the spatial overtones of these discussions, they remain far more concerned with the intricacies of spatial interaction than implied by Bloch. See, for example, Giddens on the importance of contextuality and locale and Gregory on "situational ecology": Giddens, *Constitution of Society*, pp. 355–372; Gregory, "Presences and absences," pp. 194–195.

31 Giddens, "A Reply," p. 276; Gregory, "Presences and absences," p. 194.

32 Roger Chartier, "Le Monde come représentation," *AESC*, 44 (1989), 1508–1509.

33 See, for example, Derek Gregory, "Human agency and human geography," *Transactions of the Institute of British Geographers*, n.s. 6, 1 (1981), 1–18; Anne Buttimer, "Charism and context: the challenge of *la géographie humaine*," in David Ley and Marwyn S. Samuels, *Humanistic Geography, Prospects and Problems* (Chicago: Maaroufa Press, 1978), pp. 58–76; Vincent Berdoulay, "The Vidal–Durkheim debate" in Ley and Samuels, pp. 77–90.

Kevin Archer's recent article placing Vidal within a Lamarckian tradition appears to have some potential, but overstates its point and tries to extract a theoretical sophistication and unity from Vidal's work which does not appear to have been there. Archer also underestimates the importance of contextual and institutional factors in shaping Vidal's arguments. Kevin Archer, "Regions as social organisms: the Lamarckian characteristics of Vidal de la Blache's regional geography," *Annals of the Association of American Geographers*, 83 (1993), 498–514.

Much closer to the mark is the work on Vidal by Marie-Claire Robic. Marie-Claire Robic, "L'Invention de la 'géographie humaine' au tournant des années 1900: les Vidalians et l'écologie," in Paul Claval, ed., *Autour de Vidal de la Blache: La formation de l'école française de géographie* (Paris: CNRS, 1993), pp. 137–147; Marie-Claire Robic, "La Conception de la géographie humaine chez Vidal de la Blache d'après les *Principes de géographie humaine*," *Les Cahiers de Fontenay*, 4 (1976), 1–76.

Index of Names

Gottl, Friederick, 40
Gourou, Pierre, 235 n.46
Gradmann, Robert, 138, 143
Granet, Marcel, 98, 104, 106, 128, 203 n.3
  Bloch's work compared to, 84, 85, 157, 216
    n.38
  at ENS and Fondation Thiers, 30, 75, 77,
    178, 204 n.13
Grébaut, Eugène, 26
Gregory, Derek, 181, 243 n.30
Grenier, Albert, 106, 109
Guignebert, Charles, 117, 192 n.22
Guiraud, Paul, 26

Halbwachs, Maurice, 109, 119, 128, 135, 141,
  178
  Bloch's reviews of, 121, 128–9, 131–2, 177,
    182, 221 n.96
  cf. *La Société féodale*, 156, 159
  review of *Les Rois thaumaturges*, 116–17,
    216 n.38, 217 n.52
  Annales and, 102–3
  at University of Strasbourg, 10, 95, 97–8,
    209 n.13, 222 n.2
Halecki, O. de, 101
Halévy, Elie, 23, 24, 50
Hallays, André, 107
Halphen, Louis, 148, 157–8
Hardy, Georges, 30
Harsin, Paul, 237 n.66
Hauser, Henri, 40, 45, 136, 150, 155, 235
  n.46
  and *Annales*, 102, 104, 106
Heilbron, Johan, 127–8, 131, 150, 222 n.111,
  230 n.88, 231 n.93
Héring, J., 209 n.21
Herr, Lucien, 75, 77, 203 n.3
Herriot, Edouard, 107
Hertz, Robert, 98, 128
Hill, Alette Olin, 219 n.74
Hill, Boyd, 219 n.74
Hill, David, 98
Himly, Auguste, 24–5
Hitler, Alfred, 11
Hubert, Henri, 24, 75, 128, 150, 151, 193
  n.41, 235 n.46
Hubert, René, 128
Huizinga, Johann, 116
Huntington, Ellsworth, 201 n.52
Hütter, Jean-Paul, 161, 235 n.46

Jacob, E. F., 216 n.39
Jacobsen, Lis, 162, 235 n.46, 236 n.49
Jaurès, Jean, 21, 38, 203 n.3
Jeanmaire, Henri, 193 n.35
Joffre, Joseph, 95
Jullian, Camille, 14, 34, 39, 109, 126

Karady, Victor 191 n.13
Kleinclausz, A., 34, 204 n.19
Koht, Halvdan, 187 n.33
Kötzschke, Rudolf, 9

La Senne, Ernest, 75
La Vieille, François, 30
Labouret, Henri, 104, 162–3, 235 n.46
Labrousse, Ernest, 155
Lacombe, Paul, 46, 50, 99, 193 n.41, 195 n.8
  and Bloch, 52, 53
  De l'histoire, 41, 42, 55
  at Société Française de Philosophie, 48–9,
    50, 197 n.43
Lacoste, Georges de, 75
Lalande, André, 51–2
Lambert, Edouard, 123
Lamprecht, Karl, 10, 40, 193 n.41
Langlois, Charles-Victor, 31, 180, 193 n.41
  Bloch and, 87–8, 125, 164, 165–8, 169, 171,
    179–80, 237 n.68
  at University of Paris, 26, 28–9, 37, 47–8,
    99
Lanson, Gustave, 36, 37, 93, 107, 117
Lapie, Paul, 128
Lapparent, Albert de, 206 n.44
Laskine, Edmond, 216 n.38
Lasserre, Pierre, 37
Laurent-Vibert, 30
Laval, Pierre, 11
Lavisse, Ernest, 31, 41, 64, 99, 100, 180, 189
  n.3
  at University of Paris, 21–2, 26, 37, 192
    n.21
Layton, Walter, 87
Le Bras, Gabriel, 160, 162, 209 n.21, 227 n.43,
  235 n.46
Le Chatelier, Alfred, 23
Le Goff, Jacques, 214 n.8
Le Lannou, Maurice, 140
Le Play, Frédéric, 23, 59, 198 n.4
Leclerc, Max, 102–3, 211 n.43, 223 n.11
Lefebvre, Georges, 160, 213 n.66, 235 n.46
Lefebvre, Théodore, 140
Lefebvre des Noettes, Commandant Richard,
  154, 232 n.12
Legras, Henri, 75
Leland, Waldo, 101
Léon, Paul, 191 n.20
Leroux, Emmanuel, 203 n.4
Letanconnaux, G., 106
Lévy-Bruhl, Henri, 77
Lévy-Bruhl, Lucien, 11, 24, 75, 98
  Les Rois thaumaturges and, 111, 116, 215
    n.36, 216 n.38
Liard, Louis, 31, 32–3, 37
Lipson, Ephraim, 101

Subject Index

Cambridge Studies in Historical Geography

Printed in the United States
By Bookmasters